RESEARCH ON SOIL EROSION

Edited by **Danilo Godone, Silvia Stanchi**

Research on Soil Erosion

http://dx.doi.org/10.5772/3402
Edited by Danilo Godone, Silvia Stanchi

Contributors

Tom Mueller, Adam Pike, Eduardo Rienzi, Surendran Neelakantan, Blazan Mijatovic, Tasos Karathanasis, Marcos Rodrigues, Zicheng Zheng, Shuqin He, ALi Haghizadeh, Demetrio Antonio Zema, Miroslav Dumbrovsky, Nuray Misir, Mehmet Misir, Charles Igwe, Adélia Nunes, Luciano Lourenço, António Bento Gonçalves, António Vieira, Adelaide Munodawafa

Published by InTech

Janeza Trdine 9, 51000 Rijeka, Croatia

Copyright © 2012 InTech

Notice

Publishing Process Manager Dragana Manestar
Typesetting InTech DTP Team
Cover InTech Design Team

Additional hard copies can be obtained from orders@intechopen.com

Research on Soil Erosion, Edited by Danilo Godone, Silvia Stanchi
p. cm.
ISBN 978-953-51-0839-9

For EU product safety concerns:
IN TECH d.o.o., Prolaz Marije Krucifikse Kozulić 3, 51000 Rijeka, Croatia,
info@intechopen.com or visit our website at intechopen.com.

Meet the editors

Danilo Godone holds a PhD in "Agriculture, Forest and Food Sciences", doctorate's topic was cryosphere's phenomena monitoring by geomatic methodologies. Currently he is a PostDoc grant holder, at Turin University, studying geomatic contribution in land management and analysis. His main research interests are, however, landslide, glacier and, more generally, natural disasters monitoring. During his activities he has developed skills in GIS, land surveying with GPS or Laser Scanners and data analysis by R programming. He is a member of NATRISK - Research Centre on Natural Risks in Mountain and Hilly Environments, in the same University. He acts as a freelance consultant, in the same topics, for other research bodies, training agencies and professionals, too. Outside his research and professional activities, he carries on his passion for mountain by practising hiking, snowshoeing and climbing. Additionally, he is a, martial arts and self-defence, instructor.

Silvia Stanchi holds a PhD in "Agriculture, Forest and Food Sciences", with a focus on fractal theory approaches for the study of soil aggregation and structure. Currently, she is a PostDoc at Turin University working on soil hazards and vulnerability in mountain areas. Silvia Stanchi is a member of NATRISK - Research Centre on Natural Risks in Mountain and Hilly Environments, in the same University. She is actively involved in outreach and soil awareness raising activities targeting kids and the large public, and in international training activities focusing on mountain development and management.

Contents

Preface

Soil loss for erosion is a natural phenomenon in soil dynamics, influenced by climate, soil intrinsic properties, and morphology, that can both trigger and enhance the process. Anthropic activities, like inappropriate agricultural practices, deforestation, overgrazing, forest fires and construction activities, may exert a remarkable impact on erosion processes or, on the other hand, contribute to soil erosion mitigation through a sustainable management of natural resources.

The phenomena is also exacerbated by socio economical dynamics and land management policies in developing or emerging countries, by global change and the interactions between these planetary issues.

In this book we collected a series of papers on erosion, focusing on a variety of aspects of the erosion phenomena.

The book is the continuation of previously published "Soil Erosion Studies"; it is organized in a unique section, starting with a chapter by Zema et al. concerning the comparison between two watersheds and the application of runoff model results. The experiments are carried out in two different European areas, southern Italy and central Belgium. The second one, written by Dumbrovsky focuses on soil erosion and flood control systems for territorial planning and natural hazard prevention.

Pike e al., in the third one discuss about the application of terrain analysis and survey, obtained by various methodologies and sources, in erosion modelling, particularly applying statistical analyses in determining erosion channel formation.

The fourth chapter, written by Lourenço et al. focuses on the enhancement of erosion processes after wildfires in Mediterranean environments (two catchments in Portugal), investigated through field surveys and GIS applications.

In chapter five Misir and Misir introduce the topic of forest management and its close relation to soil erosion and its mitigation through sustainable practices. Similarly Zheng and He, in chapter six, deal with agricultural practices in a critical area of the world, concerning erosion, the Chinese Loess Plateau. They analyse, by experimental plots the effects of splash erosion under different tillage practices.

Munodawafa in chapter sevenpresents the effects of land management practices on agricultural soils, determining the losses of fertility that can threaten seriously food security in Zimbabwe.

Chapter eight, by Igwe, focuses on the main effects of land management and land use changes on soil erosion in a study area in Nigeria, suggesting some possible protection strategies.

The book ends with Haghizadeh's chapter which treats the topic of daily flow model introducing the complete process from data collection to calibration and from data analysis to results comments. The study is focused on river basin in Iran and its purposes extend, beyond soil erosion, to flood prevention.

Danilo Godone, PhD
Silvia Stanchi, PhD

Turin University
NATRISK - Research Centre on Natural Risks in
Mountain and Hilly Environments
Grugliasco (TO), Italy

Section 1

Soil Erosion

Chapter 1

Prediction of Surface Runoff and Soil Erosion at Watershed Scale: Analysis of the AnnAGNPS Model in Different Environmental Conditions

Demetrio Antonio Zema, Giuseppe Bombino,
Pietro Denisi, Feliciana Licciardello and
Santo Marcello Zimbone

Additional information is available at the end of the chapter

http://dx.doi.org/10.5772/50427

1. Introduction

Negative effects of surface runoff and soil erosion in watersheds can be controlled and mitigated through hydrological models. Moreover, they are suitable to simulate various combinations of different scenarios of land and water management in a watershed and therefore they are useful for comparative analysis of different options and as a guide to what Best Management Practices (BMPs) can be adopted to minimize pollution from point and non-point sources (Shrestha et al., 2006).

Continuous simulation models (e.g. AnnAGNPS, WEPP, SWAT, etc.) provide great advantages over event-based models as they allow watersheds and their response to be studied over a longer time period in an integrated way. Nowadays, several continuous watershed-scale erosion models are available: however, relatively little validation of their performance under varying climatic and land use conditions has been carried out. The latter is an essential step before a model can be reliably applied.

The AnnAGNPS (Annualized Agricultural Non-Point Source) model (Geter and Theurer, 1998; Bingner and Theurer, 2001) is among the distributed models developed to evaluate the continuous hydrologic and water quality responses of watersheds. Many major hydrologic concepts of the single-event AGNPS model (Young et al., 1987) have been updated through the continuous simulation modeling of watershed physical processes (Baginska et al., 2003).

AnnAGNPS has been implemented to assess runoff water amount and quality as well as sediment yield in small to large monitored watersheds (ranging from 0.32 to 2500 km^2) under different environmental conditions. Such applications were frequently coupled with calibration/validation trials. Poor AnnAGNPS predictions of sediment and nutrient loads were achieved in a Georgia watershed, covered by both extensive forest and riparian conditions and attributed this to the defective data input used with the model (Suttles et al., 2003). Moderate accuracy in model simulation of phosphorous and nitrogen processes was also highlighted by model applications in two small watersheds located in the Mississippi Delta (Yuan et al., 2005) and in the Sydney region (Baginska et al., 2003). The capability of the model (coupled to the BATHTUB eutrophication reservoirs model) in simulating nutrients load variations in response to land use changes in a Kansas large reservoir was pointed out by Wang et al. (2005).

In applications to a small Mississippi watershed reported by Yuan et al. (2001, 2005), AnnAGNPS adequately predicted long-term monthly and annual runoff and sediment yield and predicted and observed runoff from individual events were reasonably close, achieving coefficients of determination r^2 and efficiency E (Nash and Sutcliffe, 1970) equal to 0.94 and 0.91 respectively). In a small Australian watershed, mainly covered by farming and residential land uses, acceptable model predictions (E = 0.82) were assessed for runoff at event scale after the calibration of hydrological parameters Baginska et al. (2003).

More recently AnnAGNPS was implemented at a small Nepalese watershed, mainly forested and cultivated, where the need of calibration for satisfactory runoff predictions was shown. Despite the calibration process, peak flow and sediment yield evaluation resulted in a much lower accuracy (Shrestha et al., 2006). The prediction performance of AnnAGNPS in a 48-km^2 watershed located in Kauai Island (Hawaii, USA) was considered good for monthly runoff predictions and poor on a daily basis (Poliakov et al., 2007). Calibration/validation tests in two small watersheds in S. Lucia Island (British West Indies) (agricultural and forested respectively) suggested that AnnAGNPS could be used under the conditions tested tested (Sarangi et al., 2007). In an agricultural river basin (374 km^2) of Czech Republic suspended load following short duration intensive rainfall events was accurately predicted by the AnnAGNPS model; there the model was not suitable for continuous simulation in large river basins with a high proportion of subsurface runoff (Kliment et al., 2008). In a 63-km^2 watershed in Malaysia (tropical region which sometimes experiences heavy rainfall runoff) was well predicted while results with respect to sediment load were moderate (Shamshad et al., 2008).

Some applications in Spanish catchments covered by olive orchards showed the sensitivity of AnnAGNPS to different temporal scales in modeling runoff and sediment yield under different management systems (Aguilar and Polo, 2005) and the model applicability to predict runoff and sediment at event and monthly scales after calibration (Taguas et al., 2009). A calibration/validation exercise using a 10-year hydrological database in 53-km^2 watershed in Ontario (Canada) highlighted that adjustments of the monthly curve number values and of the RUSLE parameters are relevant to improve the hydrology and sediment components of AnnAGNPS, especially during winter and early spring periods (Das et al., 2009). A good model performance was obtained in terms of runoff and erosion prediction after calibration/

validation processes in a 136-km^2 agricultural watershed in south-central Kansas; total phosphorus predictions were instead good only for the calibration period (Parajuli et al., 2009). Finally, a poor model performance in simulating agricultural pollution by nitrogen, phosphorus and sediment was obtained in a 16.97-km^2 watershed located in North Dakota (USA), mainly due to the large size of the study area and the high variability in land use and management practices (Lyndon et al., 2010).

Thus, the results of AnnAGNPS evaluations that have hitherto been carried out are generally promising. At the same time it can be noticed that model performance is variable and the boundary conditions under which the model may be successfully used for runoff and sediment yield prediction have not been well defined.

2. Aim of the work

In order to consolidate use of the AnnAGNPS model in different climatic and geomorphologic conditions, this investigation has verified model prediction capability of surface runoff, peak flow and sediment yield in two small European watersheds under climate conditions typical of the semi-arid (Cannata watershed, southern Italy) and humid-temperate (Ganspoel watershed, central Belgium) environments respectively. Through this work we have investigated to what extent AnnAGNPS may be expected to provide usable results in environmental conditions outside of research watersheds, where sometimes the necessary data for model calibration and validation are not available.

3. The AnnAGNPS model

AnnAGNPS is a distributed parameter, physically based, continuous simulation, daily time step model, developed initially in 1998 through a partnering project between the USDA Agricultural Research Service (ARS) and the Natural Resources Conservation Service (NRCS). The model simulates runoff, sediment, nutrients and pesticides leaving the land surface and shallow subsurface and transported through the channel system to the watershed outlet, with output available on an event, monthly and annual scale. Required inputs for model implementation include climate data, watershed physical information, as well as crop and other land uses as well as irrigation management data.

Because of the continuous nature of AnnAGNPS, climate information, which includes daily precipitation, maximum and minimum temperatures, dew point temperatures, sky cover and wind speed, is necessary to take into account temporal weather variations. The spatial variability of soils, land use, topography and climatic conditions can be accounted for by dividing the watershed into user-specified homogeneous drainage areas. The basic components of the model include hydrology, sedimentation and chemical transport.

The SCS curve number technique (USDA-SCS, 1972) is used within the AnnAGNPS hydrologic submodel to determine the surface runoff on the basis of a continuous soil moisture balance. AnnAGNPS only requires initial values of curve number (CN) for antecedent moisture condition AMC-II, because the model updates the hydrologic soil conditions on the basis of the daily soil moisture balance and according to the crop cycle.

The peak flow is determined using the extended TR-55 method (Cronshey and Theurer, 1998). This method is a modification of the original NCRS-TR-55 technology (USDA-NRCS, 1986), which is considered as a robust empirical approach suitable for wide variety of conditions including those where input data might be limited as in the experimental watershed (Polyakov et al., 2007).

The AnnAGNPS erosion component simulates storm events on a daily basis for sheet and rill erosion based on the RUSLE method (Revised Universal Soil Loss Equation, version 1.5, Renard et al., 1997). The HUSLE (Hydrogeomorphic Universal Soil Loss Equation, Theurer and Clarke, 1991) is used to simulate the total sediment volume delivered from the field to the channel after sediment deposition.

The sediment routing component simulates sheet and rill sediment deposition in five particle size classes (clay, silt, sand and small and large aggregates) on the basis of density and fall velocity of the particles and then routes sediment separately through the channel network to the watershed outlet as a function of sediment transport capacity (calculated by the Bagnold equation; Bagnold, 1966). A key assumption is that the aggregates break up into their primary particles once they enter the stream channel.

For the chemical component of the model, dissolved and adsorbed sediment predictions are assessed for each cell by a mass balance approach. Algorithms for nutrient (nitrogen, phosphorous and organic carbon) and pesticide dynamics are largely similar to the EPIC (Williams et al., 1984) and GLEAMS (Leonard et al., 1987) models.

More details on the theoretical background of AnnAGNPS are reported by Bingner and Theurer (2005).

4. Description of the Experimental Watersheds

The input data utilised for AnnAGNPS implementation in the Cannata watershed was collected during a proper monitoring campaign providing topographic, soil and land use data as well as 7-year hydrological observations.

For model verification in the Ganspoel watershed the input database was drawn from the works by Steegen et al., 2001 and Van Oost et al., 2005. Compared to the Cannata watershed, this experimental database reported less geomorphological information; moreover, the hydrological observations were related only to a 2-year period: thus this study case represents a typical "data-poor environment" (Merritt et al., 2003).

4.1. Cannata watershed

4.1.1. Geomorphological information

The Cannata watershed, located in eastern Sicily, southern Italy (outlet coordinates 37 53'N, 14 46'E), is a mountainous tributary, ephemeral in flow, of the Flascio River (Figure 1).

The watershed covers about 1.3 km^2 between 903 m and 1270 m above mean sea level with an average land slope of 21%. The longest channel pathway is about 2.4 km, with an average slope of about 12% (Figure 2). The Kirpich concentration time is 0.29 h.

Figure 1. View of the Cannata watershed in proximity of its outlet.

In a survey conducted at the start of experimental campaign, five different soil textures (clay, loam, loam-clay, loam-sand and loam-sand-clay) were recognized on 57 topsoil samples; clay-loam (USDA classification) resulted as the dominant texture. The soil saturated hydraulic conductivity, measured by a Guelph permeameter, resulted in the range 0.2 to 17.6 mm h^{-1}.

Continuous monitoring of land use has highlighted the prevalence of pasture areas (ranging between 87% and 92% of the watershed area) with different vegetation complexes (up to 15 species) and ground covers. Four soil cover situations can be distinguished: a high-density herbaceous vegetation (eventually subjected to tillage operations), a medium-density herbaceous vegetation, sparse shrubs and cultivated winter wheat with a wheat-fallow rotation. More detailed information about the watershed characteristics and the monitoring equipment were reported previously (Licciardello and Zimbone, 2002).

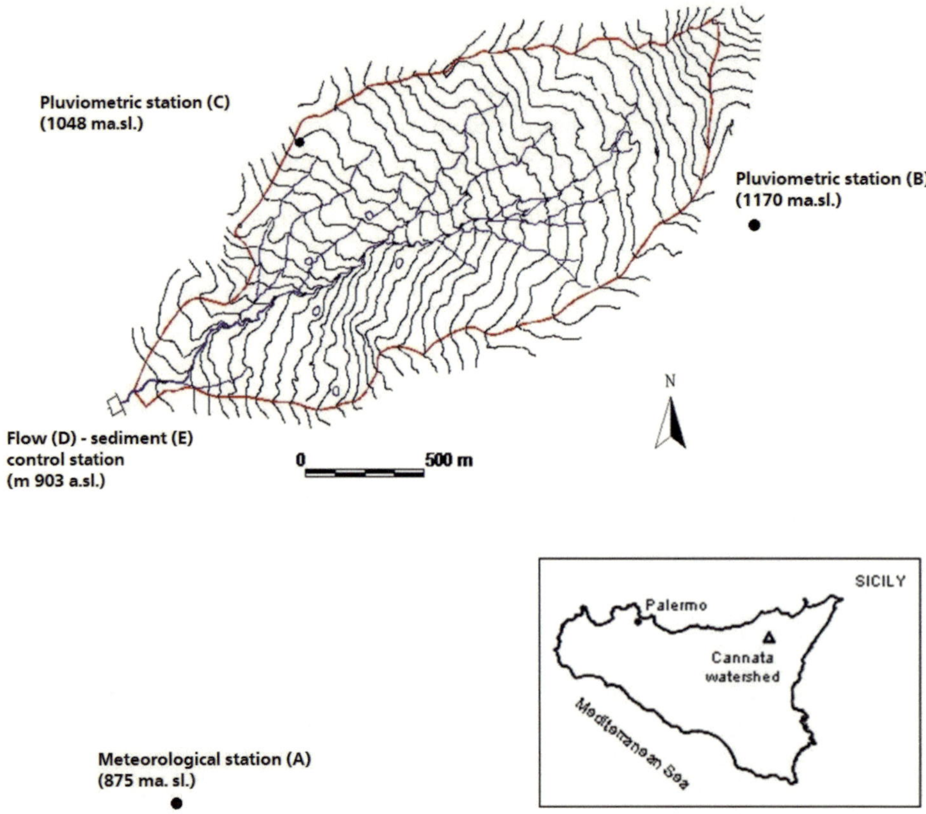

Figure 2. Location, contour map and hydrographic network of the Cannata watershed.

4.1.2. The hydrological database

In the monitoring period of 1996 to 2003 the hydrological observations were collected utilising the following equipment (Figure 2): a meteorological station (A, located outside of the watershed) recording rainfall, air temperature, wind, solar radiation and pan evaporation; two pluviometric stations (B and C); and a hydrometrograph (D) connected to a runoff water automatic sampler (E) for the measurement of sediment concentration in the flow.

In the observation period yearly rainfall between 541 and 846 mm (mainly concentrated from September to March) was recorded at the station A, with a mean and standard deviation (SD) of 662 and 134 mm respectively. The corresponding yearly runoff was in the range 30.7 to 365.8 mm, with a mean of 105.3 mm and SD of 100 mm. The coefficient of yearly runoff, calculated as the ratio between total runoff and total rainfall as recorded by station A, varied between 5% and 41%, with a mean and SD of 15% and 75% respectively. Occasional high differences in recorded rainfall events between the three gauges were found; as expected, rainfall spatial variability decreased on a monthly and yearly basis.

At event scale, rainfall depths over 6.8 mm gave runoff volumes higher than 1 mm; the maximum runoff volume and discharge recorded in the observation period were 159.6 mm and 3.4 m^3 s^{-1} (2.6 l s^{-1} km^{-2}) respectively. Twenty-four erosive events were sampled with a suspended sediment concentration between 0.1 and 9.2 g l^{-1}; the maximum event sediment yield (estimated on the basis of runoff volume and suspended sediment concentration in the flow) was 283 Mg (2168.4 kg ha^{-1}).

4.2. Ganspoel watershed

4.2.1. Geomorphological information

The Ganspoel watershed (outlet coordinates 50 48'N, 4 35'E), located in central Belgium, covers 1.15 km^2 between 60 m and 100 m a.s.l. with an average slope of about 10%, but which can locally exceed 25%. A dense network of dry channels characterizes the area (Figure 3). The topography of the area is formed in sandy deposits overlain by a loess layer that was deposited during the latest glacial period. Soils are therefore dominantly loess-derived luvisols, with their physical parameters related much more to land use than to soil texture (Van Oost et al., 2005).

Top soils have a very high silt percentage (on the average 75%) and moderate clay and sand content (on the average 11% and 14% respectively) (Van Oost et al., 2005).

The watershed land use is mainly agricultural. Forested (5%) and pasture (4%) zones cover the steep slopes as well as some of the thalweg areas. A built-up zone is located in northwestern part of the Ganspoel watershed and represents 9% of its area (Steegen et al., 2001). The main

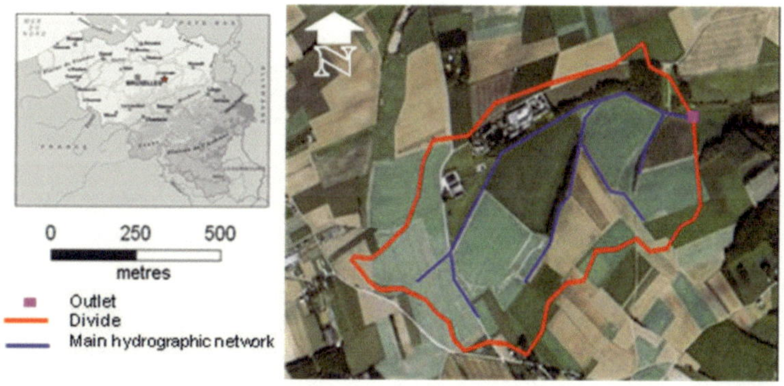

Figure 3. Location and aerial view of the Ganspoel watershed.

4.2.2. The hydrological database

The climate of central Belgium shows relatively cool summers and mild winters resulting in an average annual temperature of 11 C. Annual precipitation varies normally between 700 and 800 mm year^{-1} and is well distributed over the year. High intensity rainfall events occur mainly in spring and summer: such thunderstorms may reach peak rainfall intensities of ca. 70 mm h^{-1} while total rainfall amounts may amount to 40 mm, exceeding rarely 60 mm.

The hydrological database was collected during a recording period of about 2 years (May 1997-February 1999). The rainfall and flow/sediment measurement station was located at the outlet of the watershed. The rainfall events were recorded by a tipping-bucket rain gauge (logging interval equal to 1 minute with 0.5-mm tips). Water depths were continuously measured with a time interval of 2 minutes and an accuracy of 2 mm by a San Dimas flume equipped with a flowmeter, using a submerged probe level sensor. Water discharge was then calculated by a constant relationship between water depth and discharge. The suspended sediment concentration, measured by an automated water sampler which a flow-proportional sampling rate (every 30 m^3 runoff), was determined by oven-drying every sample at 105 C for 24 hours.

Seventeen runoff events, corresponding to rainfall depths in the range 5.5-57.5 mm, were adequately sampled (Table 1). The sampled events concerned generally low runoff volumes (15 with runoff depths lower than 2 mm), but the most intense event (13-14 September 1998) produced a runoff volume of 9.5 mm. Event-based sediment yields were in the range 2 to 604 kg ha^{-1} (Table 1). Ten other events were not taken into account because of inadequate sampling.

Event	Rainfall depth (mm)	Rainfall duration (h)	Runoff volume (mm)	Runoff coefficient (%)	Peak flow (m^3 s^{-1})	Sediment yield (Mg)	Sediment yield (kg ha^{-1})
19/05/1997	8.0	0.4	0.22	2.8	0.103	8.2	70.1
21/05/1997	6.5	8.4	0.13	2.0	0.056	2.7	23.3
11/07/1997	13.0	0.6	1.97	15.2	0.862	40.9	349.7
14/07/1997	5.5	0.6	0.37	6.7	0.181	4.4	37.6
17-18/07/1997	21.5	8.4	0.35	1.6	0.050	3.6	30.8
25/12/1997	6.5	1.0	0.09	1.4	0.043	0.2	2.1
05/01/1998	8.0	4.2	0.23	2.9	0.051	0.5	4.5
28/04/1998	11.0	1.4	0.14	1.3	0.037	0.2	1.8
05/06/1998*	10.5	3.3	0.002	0.02	0.003	-	-
06/06/1998*	29.5	32.8	13.08	44.3	1.827	-	-
11/06/1998*	16.5	21.4	3.68	22.3	0.389	-	-
22/08/1998*	36.5	47.2	0.93	2.5	0.046	-	-
26/08/1998	5.5	8.4	0.39	7.1	0.064	1.9	16.2
08-09/09/1998	24.5	1.5	0.45	1.8	0.067	1.3	11.1
13-14/09/1998	57.5	19.1	8.86	15.4	1.017	66.1	565.2

Event	Rainfall depth (mm)	Rainfall duration (h)	Runoff volume (mm)	Runoff coefficient (%)	Peak flow ($m^3 s^{-1}$)	Sediment yield (Mg)	Sediment yield ($kg ha^{-1}$)
31/10-01/11/1998	25.0	19.3	1.67	6.7	0.064	6.9	58.9
14/11/1998	15.5	14.4	0.71	4.6	0.032	0.7	6.1
29/11/1998	18.5	19.9	0.56	3.0	0.025	1.4	12.0
07/12/1998*	7.0	60.8	0.93	13.3	0.026	-	-
19/12/1998*	4.5	5.7	0.27	6.0	0.033	-	-
07/01/1998*	28.0	51.5	1.80	6.4	0.061	-	-
16-17/01/1999	14.5	21.0	0.94	6.5	0.033	2.6	21.8
25/01/1999*	21.5	49.5	1.61	7.5	0.788	-	-
28/01/1999	8.0	3.8	0.71	8.9	0.046	3.0	25.6
07/02/1999	6.5	12.0	0.30	4.6	0.029	0.5	4.7
21/02/1999*	8.0	49.5	2.36	29.5	0.768	-	-
01/03/1999*	6.0	8.1	1.29	21.5	0.777	-	-

* Event not taken into account, because of inadequate sampling (see Van Oost et al., 2005 for more details).

Table 1. Main characteristics of the observed events used for the AnnAGNPS model implementation at the Ganspoel watershed (Ganspoel database, 2007).

5. Model implementation

The watershed discretization into homogeneous drainage areas ("cells") and the hydrographic network segmentation into channels ("reaches") were performed for both watersheds using the GIS interface incorporated into AnnAGNPS.

The geometry and the density of the drainage network were modeled by setting the Critical Source Area (CSA) to 1.25 ha and the Minimum Source Channel Length (MSCL) to 100 m for the Cannata watershed, which allowed a suitable representation of the same watershed in a previous study (Licciardello et al., 2006). Such values were decreased to 0.5 ha and 50 m respectively for the Ganspoel watershed, because of its higher land use heterogeneity (Nearing et al., 2005). The Cannata watershed resulted in 78 cells and 32 reaches (Figure 4a), while the Ganspoel watershed in 155 cells and 65 reaches (Figure 4b).

The elevation GIS layer was arranged by digitizing contour lines every 2 m on a 5-m resolution DEM; land use and soil input data were derived from 25-m resolution GIS maps. The morphologic parameters (i.e., cell slope length and steepness) as well as the dominant land uses and soil types were directly associated with each drainage area by means of the GIS interface.

Meteorological and pluviometric input data were properly arranged by the AnnGNPS weather subroutines. For the Cannata watershed daily values of maximum and minimum air temperatures, relative humidity, solar radiation and wind velocity were measured at the meteorological station within the watershed. Daily rainfall input data were derived from records provided by the three working rain gauges in the different periods and input to each drainage area by applying the Thiessen polygon method, except when only the rainfall recorded at a single station was available (Figure 2). For the Ganspoel watershed, as no meteorological information (except for rainfalls) was provided in the database, air temperature, relative humidity and wind velocity data were collected at the nearest meteorological station (Bruxelles, 50 54'N, 4 30'E, about 13 km far from the watershed outlet). Solar radiation was evaluated by the Hargreaves' formula. For both watersheds daily values of dew point temperature were calculated on the basis of air temperature and humidity.

Figure 4. Layouts of the Cannata (left) and Ganspoel (right) watershed discretisation by the AnnAGNPS model.

To allow the model to adjust the initial soil water storage terms, the first two years were appended to the beginning of the precipitation and meteorological dataset. The initial values of CN, unique throughout the whole simulation period, were initially derived from the standard procedure set by the USDA Soil Conservation Service (Table 2).

Table 3 shows the values or range of the RUSLE parameters set utilised by the erosive submodel. The average annual rainfall factor (R), its cumulative percentages for 24 series of 15-day periods in a year and the soil erodibility factor (K) were determined according to guidelines by Wischmeier and Smith (1978), the latter on the basis of a field survey of soil hydrological characteristics (Indelicato, 1997; Steegen et al., 2001; Van Oost et al., 2005).

In the Cannata watershed, for each of the five soil textures, a uniform soil profile was modeled up to 1500 mm by averaging the required physical characteristics from the field samples. Soil wilting point and field capacity were derived from the experimental dataset. The whole Ganspoel watershed was modelled assuming a unique soil type (silt loam) up to a depth of 1000 mm. Values of soil wilting point and field capacity, not available from the Ganspoel dataset, were estimated by a pedo-transfer function (Saxton et al., 1986). The values of the soil saturated hydraulic conductivity (K_{sat}, in the range 0.001-205 mm h^{-1}) was derived from the LISEM Limburg database, as these data were collected on very sim-

http://dx.doi.org/10.5772/50427

ilar soils (Takken et al., 1999; Nearing et al., 2005). Given that, as above mentioned, soil physical parameters were much more related to land use than to soil texture (Van Oost et al., 2005), six different values of K_{sat} (one for each soil land use surveyed into the watershed) were input to the model.

Parameter	Cannata			Ganspoel	
	Land use	Value default model	Value after calibration	Land use	Value
	HYDROLOGICAL SUBMODEL				
Initial curve number (CN)	Cropland	81[C]; 84[D]	75[C]; 78[D]	Cropland + urban zones	81[B]; 84[D]
	Pasture	79[C]; 84[D]	72[C]; 78[D]	Forested, meadow and fallow zones	71[B]; 78[D]
Synthetic 24-h rainfall distribution type	All	I	Ia	All	II
	EROSIVE SUBMODEL				
Sheet flow Manning's roughness coefficient ($m^{-1/3}$ s)	Pasture	0.13*	0.1*	All	0.15*
	Cropland	0.125*	0.1*		
Concentrated flow Manning's roughness coefficient ($m^{-1/3}$ s)	Pasture	0.13*	0.1*		0.04*
	Cropland	0.125*	0.1*		
Surface long-term random roughness coefficient (mm)	Pasture +cropland	32	15		16

(1) The hydrologic groups are reported in brackets
(*) According to the indications in the AGNPS user manual (Young et al., 1994) integrated with those provided by the user manual of the EUROSEM model (Morgan et al., 1998).

Table 2. Input parameters subject to calibration process of the AnnAGNPS model in the experimental watersheds.

For both waterheds vegetation cover and soil random roughness data were collected during the whole monitoring period.

Management information (crop types and rotation as well as agricultural operations) was entered in the plant/management files and modelled using the RUSLE database guidelines and database. For the crop cultivations it was necessary to modify some default parameter values such as crop planting and harvest dates as well as type and dates of agricultural operations.

The C factor was directly calculated by the model as an annual value for non-cropland and as a series of twenty-four 15-day values per year for cropland (based on prior land use, surface cover, surface roughness and soil moisture condition (AnnAGNPS, 2001; Bingner and Theurer, 2005). The practice factor (P) was always set to 1, due to the absence of significant protection measures in the watershed (Table 3).

RUSLE factor		Value or range	
		Cannata	Ganspoel
R (MJ mm ha^{-1} h^{-1} year^{-1})		1040	1496
(Mg ha^{-1} per R-factor unit)		0.39 to 0.53	0.06
LS (-)		1.72 to 4.94	0.10 to 2.29
C (-)	Cropland[a]	0.0002 to 0.042[b]; 0.0001 to 0.043[c]	0.00002 to 0.269
	Rangeland[d]	0.016[b]; 0.029[c]	0.0074
P (-)		1	

[a] Series of twenty-four 15-day period values per year (AnnAGNPS, 2001)
[b] Before calibration
[c] After calibration and for validation
[d] Annual value (AnnAGNPS, 2001).

Table 3. Values or range of the RUSLE parameters set at the experimental watersheds for the evaluation of the AnnAGNPS model.

5.1. Hydrological simulation

After processing the input parameters of the hydrological and erosive sub-models (respectively requiring the determination of the initial Curve Numbers for the USDA SCS-CN model and the calculation of the RUSLE model factors), daily values of surface runoff, peak flow and sediment yield were continuously simulated at the outlet of both watersheds by AnnAGNPS (version 3.2).

Considering that baseflow is not considered by AnnAGNPS, the surface runoff separation from baseflow was performed by the traditional manual linear method applied to observed stream flow data. Based on studies by Arnold et al. (1995) as well as Arnold and Allen (1999), these results match reasonably well with those obtained through an automated digital filter; the differences in the surface runoff component extracted by the two methods are up to 20% at yearly scale.

5.1.1. Cannata watershed

Both the hydrological and erosion components of AnnAGNPS were calibrated/validated separating the calibration and validation periods by the split-sample technique. The calibration/validation process was carried out by modifying the initial values of CN, which represent a key factor in obtaining accurate prediction of runoff and sediment yield (Yuan et al., 2001; Shrestha et al., 2006); and the most important input parameter to which the runoff is sensitive (Yuan et al., 2001; Baginska et al., 2003), besides soil (field capacity, wilting point and saturated hydraulic conductivity) as well as climate parameters (precipitation, temperature and interception).

In order to calibrate/validate the peak flows and the sediment yields, both 24-h rainfall distributions typical of a Pacific maritime climate (types I and Ia) with wet winter and dry summers (USDA-NCRS, 1986) derived by the extended TR-55 method database were used. The sediment yields were evaluated at event scale by adjusting the surface long-term random roughness coefficient (which affects the RUSLE C-factor) as well as the sheet and concentrated flow Manning's roughness coefficients (Table 3).

5.1.2. Ganspoel watershed

For simulation of surface runoff, peak flow and sediment yield events, the AnnAGNPS model run with default input parameters (Table 3). No calibration/validation processes were undertaken.

6. Model evaluation

In both the experimental watersheds surface runoff volumes and sediment yields were evaluated at the event scale; in the Cannata watershed the analysis of surface runoff was extended to the monthly and annual scale.

Model performance was assessed by qualitative and quantitative approaches. The qualitative procedure consisted of visually comparing observed and simulated values. For quantitative evaluation a range of both summary and difference measures were used (Table 4).

The summary measures utilized were the mean and standard deviation of both observed and simulated values. Given that coefficient of determination, r^2, is an insufficient and often misleading evaluation criterion, the Nash and Sutcliffe (1970) coefficient of efficiency (E) and its modified form (E_1) were also used to assess model efficiency (Table 4). In particular, E is more sensitive to extreme values, while E_1 is better suited to significant over- or underprediction by reducing the effect of squared terms (Krause et al, 2005). As suggested by the same authors, E and E_1 were integrated with the Root Mean Square Error (RMSE), which describes the difference between the observed values and the model predictions in the unit of the variable. Finally, the Coefficient of Residual Mass (CRM) was used to indicate a prevalent model over- or underestimation of the observed values (Loague and Green, 1991).

The values considered to be optimal for these criteria were 1 for r^2, E and E_1 and 0 for RMSE and CRM (Table 4). According to common practice, simulation results are considered good for values of E greater than or equal to 0.75, satisfactory for values of E between 0.75 and 0.36 and unsatisfactory for values below 0.36 (Van Liew and Garbrecht, 2003).

Coefficient or measure	Equation	Range of variability
Coefficient of determination	$r^2 = \left[\frac{\sum_{i=1}^{n} (O_i - \bar{O})(P_i - \bar{P})}{\sqrt{\sum_{i=1}^{n} (O_i - \bar{O})^2} \sqrt{\sum_{i=1}^{n} (P_i - \bar{P})^2}} \right]^2$	0 to 1
Coefficient of efficiency (Nash and Sutcliffe, 1970)	$E = 1 - \frac{\sum_{i=1}^{n} (O_i - P_i)^2}{\sum_{i=1}^{n} (O_i - \bar{O})^2}$	$-\infty$ to 1
Modified coefficient of efficiency (Willmott, 1982)	$E_1 = 1 - \frac{\sum_{i=1}^{n} \lvert O_i - P_i \rvert}{\sum_{i=1}^{n} \lvert O_i - \bar{O} \rvert}$	$-\infty$ to 1
Root Mean Square Error	$\mathrm{RMSE} = \sqrt{\frac{\sum_{i=1}^{n} (P_i - O_i)^2}{n}}$	0 to ∞
Coefficient of residual mass (Loague and Green, 1991)	$\mathrm{CRM} = \frac{\sum_{i=1}^{n} O_i - \sum_{i=1}^{n} P_i}{\sum_{i=1}^{n} O_i}$	$-\infty$ to ∞

n = number of observations.
O_i, P_i = observed and predicted values at the time step i.
$\bar{O}$ = mean of observed values.

Table 4. Coefficients and difference measures for model evaluation and their range of variability.

7. Results and discussion

7.1. Cannata watershed

7.1.1. Calibration test

The observed runoff volumes from October 1996 to December 2000 at the watershed outlet were used for model calibration at monthly and event scales; annual model performance was evaluated by utilizing observations from the years 1997 to 2000. In trying to approximate the mean and SD values of the observed runoff, the initial CNs were properly decreased both in rangeland and in cropland areas (Table 3). Table 5 shows the values of the chosen difference measures obtained for runoff at annual, monthly and event scales before and after calibration.

Values	Mean (mm)	Std. Dev. (mm)	r^2	E	E_1	RMSE (mm)	CRM
Calibration test							
Annual scale (1997 to 2000)							
Observed	78.54	40.25					
Predicted[a]	107.05	43.05	0.59	-0.13	-0.10	38.19	-0.40
Predicted[b]	77.17	39.81	0.72	0.70	0.53	6.30	0
Monthly scale (Oct. 1996 to Dec. 2000)							
Observed	7.71	15.91					
Predicted[a]	10.79	19.50	0.75	0.59	0.48	10.15	-0.40
Predicted[b]	7.70	15.98	0.78	0.77	0.61	7.61	0
Event scale (Oct. 1996 to Dec. 2000)							
Observed	0.25	2.42					
Predicted[a]	0.36	2.79	0.83	0.76	0.52	1.18	-0.40
Predicted[b]	0.25	2.36	0.85	0.84	0.64	0.96	0
Validation test							
Annual scale (Jan. 2001 to Dec. 2003)							
Observed	158.74	145.05					
Predicted[b]	108.38	80.79	0.99	0.62	0.54	72.74	0.32
Monthly scale (Jan. 2001 to Dec. 2003)							
Observed	13.23	34.43					
Predicted[b]	9.03	24.20	0.93	0.85	0.66	13.27	0.32
Event scale (Jan. 2001 to Dec. 2003)							
Observed	0.43	5.37					
Predicted[b]	0.30	4.00	0.87	0.83	0.58	2.21	0.32

[a] Default simulation
[b] Calibrated model.

Table 5. Values of the coefficients, summary and difference measures applied to runoff volumes at different time scales for calibration and validation tests at the Cannata watershed.

The simulated total runoff volume for the period of October 1996 to December 2000 (405.72 mm) was only slightly higher than the observed value (393.23 mm), showing a runoff prediction capability for long periods, which was also detected by other Authors (Yuan et al., 2001). The improvement in the annual runoff volume predictions after the calibration is due to the reduction of the cumulated volume overprediction relative to events with smaller runoff (Figure 5). In some cases, at the beginning of the wet season, runoff was generated by AnnAGNPS but not observed (Figure 6). This was probably due to the peculiarity of the hydrological processes governing runoff formation in Mediterranean regions, depending not only on catchment characteristics but also on antecedent hydrological conditions and characteristics of the rainfall events, with low runoff coefficients as a result of short-duration, high-intensity convective storms over dry soils (Latron et al., 2003).

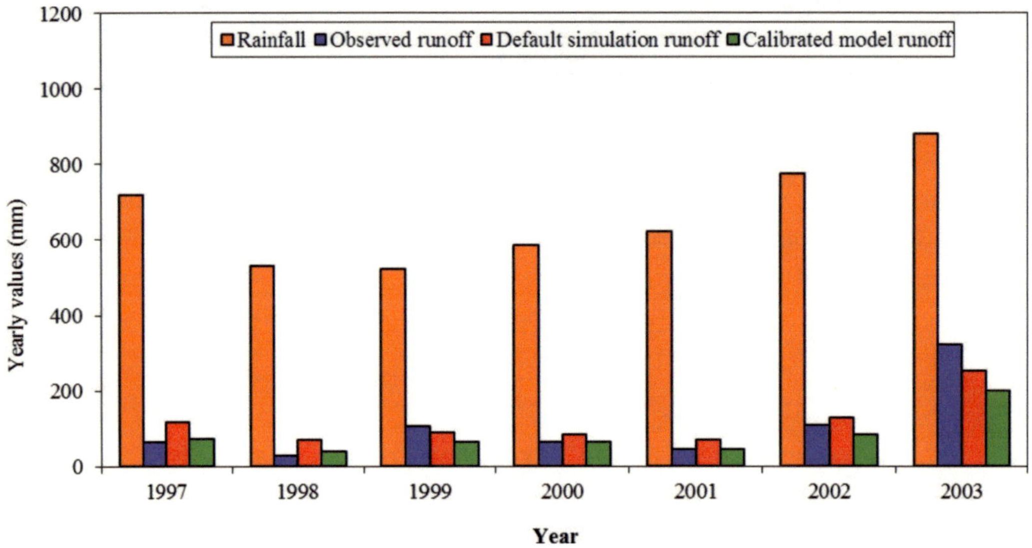

Figure 5. Comparison between observed and simulated (using default and calibrated parameters) yearly runoff volume for the years 1997 to 2003 at the Cannata watershed.

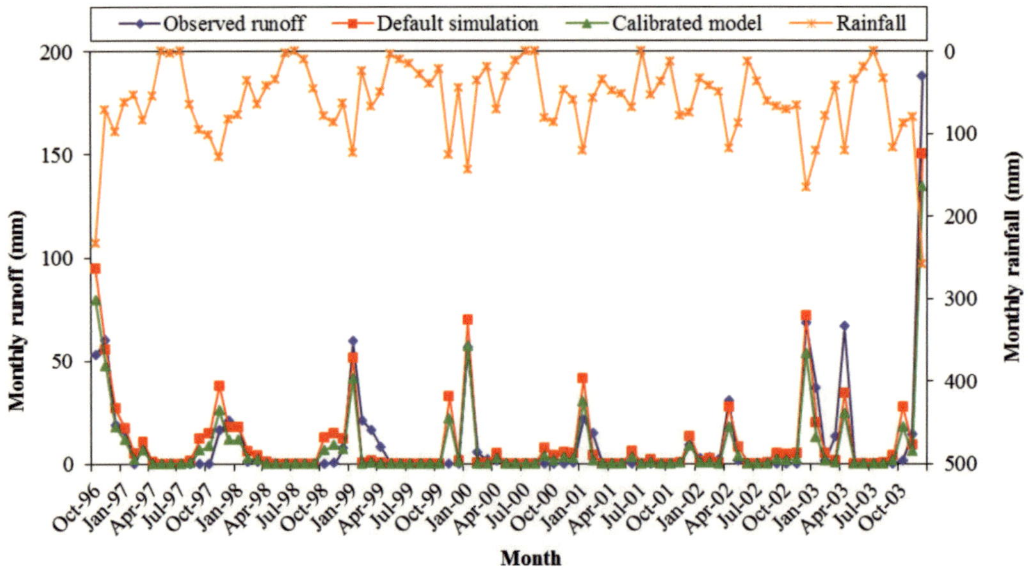

Figure 6. Comparison between observed and simulated (using default and calibrated parameters) monthly runoff volume for the whole period at the Cannata watershed.

The goodness of fit between observed and simulated runoff volumes (Figure 7) was also confirmed at the event scale by the summary measures as well as by the satisfactory values of E_1 and the low RMSE and CRM (Table 5). A similar value of E was found in the model calibration test reported by Baginska et al. (2003).

The apparent best results achieved for monthly and event-scale runoff volume predictions with respect to annual values may depend on the fact that the simulation period only represents a few years of data (four years and three years for the calibration and validation periods, respectively), while monthly and event-scale simulations provide more data for the statistics. Moreover, in Table 5, results of simulations related to the period of October to December 1996, which was very well simulated by the model, are not reported.

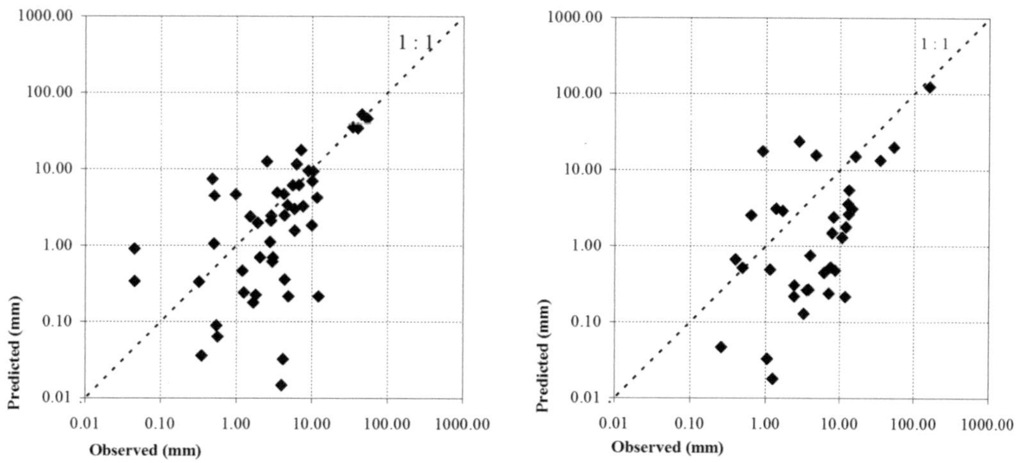

Figure 7. Comparison between observed and simulated runoff at event scale for (left) calibration and (right) validation tests at the Cannata watershed.

As expected, the coefficient E_1 is less sensitive to peaks (Krause et al., 2005) and was generally lower than E, but nevertheless satisfactory after the calibration process.

Adjustments of minimum and maximum interception evaporation (the portion of precipitation that neither runs off nor infiltrates) within the lower and upper default bounds assumed by AnnAGNPS for daily pluviometric and meteorological data did not improve the model prediction capability.

Peak flow predictions were closer to the observed values when the type Ia synthetic 24-h rainfall distribution (less intense than type I) was used. The overall model performance was satisfactory for less intense events, as shown by the E_1 coefficient (Table 6).

High values of the coefficient of determination and model efficiency (E and E_1) were found for the suspended sediment yield events observed from October 1996 to December 2000 (Figure 8) when the AnnAGNPS erosive submodel was calibrated (Table 7). By decreasing the surface long-term random roughness coefficient as well as the sheet and concentrated flow Manning's roughness coefficients for both rangeland and cropland areas, the tendency to underprediction was substantially reduced. The model response was remarkably more sensitive to the random roughness (more than 95% of the model efficiency improvement) than the Manning's coefficients adjustments (Table 3).

Values	Mean ($m^3 s^{-1}$)	Std. Dev. ($m^3 s^{-1}$)	r^2	E	E_1	RMSE ($m^3 s^{-1}$)	CRM
			Calibration test (Oct. 1996 to Dec. 2000)				
Observed	0.02	0.11					
Predicted [a]	0.03	0.33	0.57	-4.04	0.05	0.26	-1.12
Predicted [b]	0.01	0.14	0.56	0.34	0.52	0.09	0.14
			Validation test (Jan. 2001 to Dec. 2003)				
Observed	0.02	0.14					
Predicted[b]	0.02	0.23	0.66	0.05	0.51	0.14	0.11

[a] Default simulation
[b] Calibrated model.

Table 6. Values of the coefficients, summary and difference measures applied to peak flow at event scale for calibration and validation tests at the Cannata watershed.

Peak flow and sediment yield predictions were only slightly sensitive to the calibration of the hydrological submodel; the model efficiency in sediment yield prediction did not increase by adjusting either the Manning's roughness coefficient for channels or the ratio of rill to inter-rill erosion for bare soil.

Values	Mean (Mg)	Std. Dev. (Mg)	r^2	E	E_1	RMSE (Mg)	CRM
			Calibration test (Oct. 1996 to Dec. 2000)				
Observed	23.31	28.30	--	--	--	--	--
Predicted [a]	11.00	16.46	0.84	0.51	0.49	18.52	0.53
Predicted [b]	17.16	25.74	0.84	0.79	0.71	12.27	0.26
			Validation test (Jan. 2001 to Dec. 2003)				
Observed	26.17	69.13	--	--	--	--	--
Predicted [b]	32.14	81.62	0.92	0.87	0.55	24.34	-0.23

[a] Default simulation
[b] Calibrated model.

Table 7. Values of the coefficients, summary and difference measures applied to sediment yield at event scale for calibration and validation tests at the Cannata watershed.

http://dx.doi.org/10.5772/50427

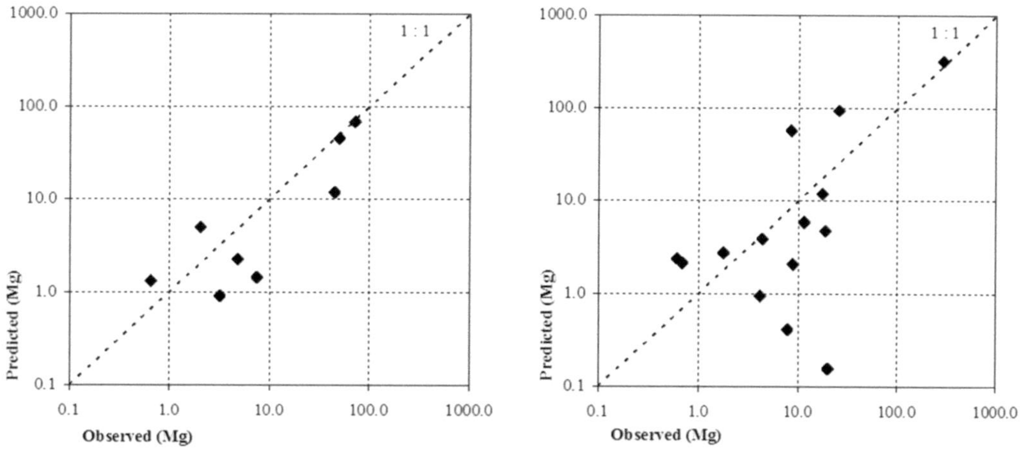

Figure 8. Comparison between observed and simulated sediment yield at event scale for (left) calibration and (right) validation tests at the Cannata watershed.

7.1.2. Validation test

The performance of the calibrated model was evaluated for the period of January 2001 to December 2003 in terms of runoff, peak flow and sediment yield.

AnnAGNPS runoff volume predictions confirmed the satisfactory model performance both at the event and annual scales and the good performance at the monthly aggregated values (Table 5). However, an underprediction was highlighted by the difference in summary measures and the values of RMSE and CRM. This tendency was mainly due to underestimation of the more significant events (Figure 7), as also found in the tests performed by Yuan et al. (2001).

The poor performance of the model in predicting extreme peak flows was confirmed in the validation period. The overall model prediction capability was unsatisfactory (Table 6), as shown by the poor value of the coefficient of efficiency (E = 0.05). A high overprediction (over 105%) for the most significant event, which occurred on 12 December 2003, is also noted.

A satisfactory model efficiency (E_1 = 0.55) and a very high coefficient of determination (r^2> 0.90) were also found for the suspended sediment yield events observed in the period of 2001 to 2003 (Table 7 and Figure 8). The satisfactory value achieved for the Nash and Sutcliffe coefficient (E = 0.87) was mainly due to the successful performance of the model for large rainfall events, in particular for the highest sediment yield, which occurred on 12 December 2003.

7.2. Ganspoel watershed

Runoff depths were in general underpredicted (see the positive value of the CRM coefficient in Table 8). The accuracy achieved for the prediction of the largest event (13-14/09/1998) gave a coefficient of determination exceeding 0.90 (Figure 9) and a model efficiency (E) of 0.89 for runoff depth (Table 8). The mean and standard deviation of simulated runoff volume depths were close to the corresponding observed values with differences lower than 12% and 16%. When the events for which zero runoff was simulated events were excluded from the analysis, the values for r^2 and E become 0.98 and 0.97 respectively. Similarly high values for the coefficient of determination were found for runoff simulations by AnnAGNPS at the event scale by Yuan et al. (2001), Shrestha et al. (2006) and Shamshad et al. (2008) and for the coefficient of determination and model efficiency by Sarangi et al. (2007). However, in these studies AnnAGNPS was calibrated before a validation was carried out.

From such outcomes it can be remarked that the AnnAGNPS model provided a generally good capability to simulate the greatest runoff event in the Ganspoel watershed, as shown by the high coefficients of efficiency (E and E_1) and determination (r^2) achieved without any a priori calibration. The latter is an important observation as it shows that, at least for significant events, adequate runoff modeling is possible without calibration provided that sufficiently detailed input data are available. The latter should not only contain land use, but also surface characteristics and soil roughness as these are important controls on runoff production. This result contrasts somewhat with that of many other studies, where the need for appropriate calibration is stressed (e.g. Refsgaard, 1997; Beven, 2006). A possible reason for this is that in many cases the available input data are less detailed than those available for the Ganspoel watersheds in terms of soil surface characteristics and coverage. The latter are important controls on runoff generation: if such data are not available, model predictions cannot be expected to be accurate without prior calibration.

The majority of the observations available in the hydrological database was of low magnitude (14 out of 17 with runoff depths lower than 1 mm); for them the model simulation accuracy was basically less accurate, achieving a mean deviation between simulations and observations of about 50%. Moreover, seven events (five of them concentrated at the end of relatively dry periods and generated by storms with a depth up to 13 mm) resulted in zero runoff simulation, even tuning the values of the initial CNs or saturated hydraulic conductivity (which represent the most important input parameters to which the runoff is sensitive (Yuan et al., 2001; Baginska et al., 2003) and setting up pre-run before the first event simulated (which is important for initial soil moisture). The AnnAGNPS model, calculating daily and sub-daily water budgets using NRCS TR-55 method coming from the SWRRB and EPIC models (Williams et al, 1984; USDA-NRCS, 1986), presumably would have adjusted the CNs to antecedent moisture condition AMC-I based on the NRCS criteria, minimising the effect of varying the CNs (Sarangi et al., 2007). The climatic characteristics of the studied watershed caused the model to produce unrealistic CN values during its initialization and, as a result, too low or no predicted runoff, as also found in various experimental applications in different climatic conditions (Polyakov et al., 2007; Sarangi et al., 2007).

Even in the Ganspoel watershed adjustments of minimum and maximum interception, as operated for model's implementation at the Cannata watershed, did not further improve the coefficients E, E_1 and r^2 calculated for runoff volume prediction.

The AnnAGNPS model provided the highest accuracy in peak flow predictions when the type "II" synthetic 24-h rainfall distribution (typical of continental climate, with cold winter and warm summer) was set in simulation tests (Figure 9). Even though statistics of observed and predicted values were of the same order of magnitude (Table 8), the low values achieved by the coefficients of efficiency (E and E_1 lower than 0.35) and conversely the high RMSE (163% of observed mean, Table 8) utilized for model evaluation confirmed the unsatisfactory prediction capability of the model for peak flow, also found elsewhere in different model tests (Shrestha et al., 2006). The model uses the extended TR-55 methods through synthetic 24-h rainfall distributions to calculate the peak flow (Cronshey and Theurer, 1998). Apparently, the latter method results is not suitable for the study area, leading to a severe underestimation of rainfall intensities and hence peak flows, a fact also noted by Shrestha et al. (2006). A prediction method that takes into account the actual patterns of rainfall intensity would be expected to provide better accuracy in peak flow estimations.

			Runoff				
Values	**Mean (mm)**	**Std. Dev. (mm)**	**r^2**	**E**	**E_1**	**RMSE (mm)**	**CRM**
Observed	1.04	2.26	-	-	-	-	-
Predicted	0.87	2.53	0.92	0.89	0.59	0.73	0.16
			Peak flow				
	Mean ($m^3 s^{-1}$)	**Std. Dev. ($m^3 s^{-1}$)**	**r^2**	**E**	**E_1**	**RMSE ($m^3 s^{-1}$)**	**CRM**
Observed	0.16	0.30	-	-	-	-	-
Predicted	0.12	0.39	0.53	0.35	0.19	0.26	0.27
			Sediment yield				
	Mean (Mg)	**Std. Dev. (Mg)**	**r^2**	**E**	**E_1**	**RMSE (Mg)**	**CRM**
Observed	8.54	17.65	-	-	-	-	-
Predicted	1.84	4.31	0.57	0.16	0.29	15.71	0.78

Table 8. Statistics concerning the AnnAGNPS simulations of 17 events at the Ganspoel watershed.

Predicted sediment yields were strongly underestimated with respect to the observed values (up to one order of magnitude in three cases); the correlation between observed and predicted values was relatively low (Table 9; Figure 9). Coefficients of efficiency (E and E_1) were close to zero and the coefficient of determination did not exceed 0.60 (Table 8). Those results were in accordance of what reported by Yuan et al. (2001), Shrestha et al. (2006), Polyakov et al. (2007) and Shamshad et al. (2008) in sediment yield modeling by AnnAGNPS.

The model tendency to strongly underpredict peak flow is probably one of the main reasons for the underestimation of erosive events and, consequently, of sediment yield (also shown by the separate comparison of deposition and erosion values for observed and simulated events, Van Oost et al., 2005), but is not the only one. Also in the case of a good estimation of the runoff volume and an overestimation of the peak flow (13-14/09/1998), the sediment yield was underestimated. Runoff alone is not adequate for erosion and sediment delivery predictions, but in the AnnAGNPS erosion sub-model it is used to estimate the delivery of the particle sizes of eroded sediment (simulated through the RUSLE model) based on runoff and peak flow.

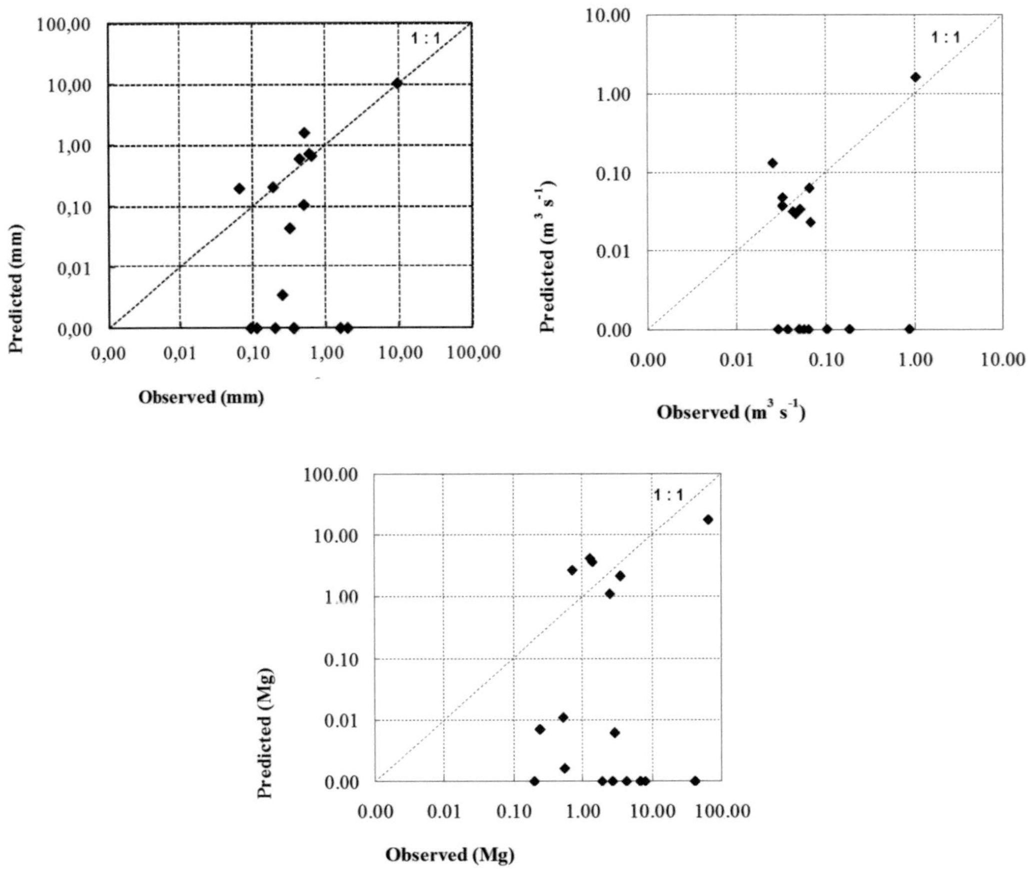

Figure 9. Comparison of 17 observed and simulated (by AnnAGNPS) events in the Ganspoel watershed, for runoff (upper left), peak flow (upper right) and sediment yield (bottom) (values are in logarithmic scale).

However, another factor that may also play a role in poor model simulations of erosion was the limited availability of input parameters. The AnnAGNPS model requires up to 100 unique parameters for runoff volume assessment and up to an additional 80 unique parameters for sediment yield prediction. As values for these parameters were not all available in the Ganspoel dataset data from the literature had to be used in some cases.

http://dx.doi.org/10.5772/50427

Date	Rainfall	Runoff volume		Peak flow		Sediment yield			
		Obs.	Sim.	Obs.	Sim.	Observed		Simulated	
	(mm)	(mm)		($m^3 s^{-1}$)		(Mg)	(kg ha^{-1})	(Mg)	(kg ha^{-1})
19/05/1997	8.0	0.22	0	0.103	0	8.20	70.09	0	0
21/05/1997	6.5	0.13	0	0.056	0	2.73	23.32	0	0
11/07/1997	13.0	1.97	0	0.862	0	40.91	349.68	0	0
14/07/1997	5.5	0.37	0	0.181	0	4.40	37.63	0	0
17-18/07/1997	21.5	0.35	0.04	0.050	0.003	3.60	30.78	2.18	18.63
25/12/1997	6.5	0.09	0.20	0.043	0.032	0.25	2.11	0.01	0.09
05/01/1998	8.0	0.23	0.21	0.051	0.034	0.53	4.53	0.01	0.09
28/04/1998	11.0	0.14	0	0.037	0	0.21	1.76	0	0
26/08/1998	5.5	0.39	0	0.064	0	1.89	16.18	0	0
08-09/09/1998	24.5	0.45	0.60	0.067	0.023	1.30	11.09	4.17	35.64
13-14/09/1998	57.5	8.86	10.55	1.017	1.629	66.13	565.19	17.66	150.94
31/10-01/11/1998	25.0	1.67	0.01	0.064	0.062	6.89	58.92	0	0
14/11/1998	15.5	0.71	0.68	0.032	0.038	0.72	6.13	2.61	22.31
29/11/1998	18.5	0.56	1.62	0.025	0.129	1.40	12.01	3.51	30.00
16-17/01/1999	14.5	0.94	0.73	0.033	0.047	2.55	21.80	1.09	9.32
28/01/1999	8.0	0.71	0.11	0.046	0.029	2.99	25.57	0	0
07/02/1999	6.5	0.30	0	0.029	0	0.55	4.69	0	0

Table 9. Main characteristics of the observed events and simulations by the AnnAGNPS model at the Ganspoel watershed.

Moreover, the following factors can explain the low correlation between observed and predicted sediment yields:

- AnnAGNPS uses the RUSLE method as the erosion sub-model. RUSLE has been developed to deliver estimates of long-term average erosion rates rather than event-based simulations. For this reason, comparison of individual events may not agree as well as long-term annual values (Shrestha et al., 2006), even in the case of adequate prediction for the most intense runoff events, as achieved in our model tests;
- we deliberately opted to evaluated the AnnAGNPS model without prior validation in order to assess its performance in cases where no data for validation are available;
- the Ganspoel watershed contains more than 80 fields (roads, buildings, forest, grassed channels and several crops with differing planting and harvesting schedules), showing difficulties for modeling of interactions between physical processes (water evapotranspiration,

interception, infiltration and runoff as well as soil detachment and transport) and water and sediment routing associated with its complexity (Nearing et al., 2005; Licciardello et al., 2009). Probably, the scale of soil property measurements within the available geomorphological database does not correspond to the discretisation scale of the Ganspoel watershed (characterized by land use heterogeneity and crop schedule complexity, as mentioned above) performed by the GIS interface of the data-intensive AnnAGNPS model.

8. Conclusion

The implementation of the AnnAGNPS in two small agricultural watersheds (Cannata, southern Italy, and Ganspoel, central Belgium) provided interesting indications about model's prediction capability of surface runoff, peak flow and sediment yield and thus about its applicability in the experimental conditions.

The study case of the Cannata watershed has highlighted a good prediction capability of runoff and erosive events, particularly for the events of highest relative magnitude (higher than 15 mm and 100 kg ha^{-1} respectively); a good accuracy has been achieved also for monthly runoff volumes simulation. The over-estimation of runoff volumes at yearly scale has been limited by setting up the initial CNs in the calibration phase, with mean differences between observed and simulated yearly values lower than 20%. Peak flow predictions have been satisfactory only for the less intense events (lower than 0.3 m^3/s); the utilisation of the different synthetic hyetographs available for the hydrologic sub-model has not hallowed to eliminate the high overestimation of the most intense peak flows. On the whole, the results provided by the analysis of this study case encourage further efforts in order to verify the model transferability to the climatic conditions typical of the semi-arid Mediterranean environment.

The evaluation of AnnAGNPS in the Ganspoel watershed has highlighted a good prediction capability only for the most intense runoff events (higher than 1 mm) in absence of calibration. The prediction capability of peak flows and sediment yields have resulted instead unsatisfactory (as also highlighted by the low coefficients of efficiency): the poor model's sediment yield predictions reflect the unreliability of simulated values of peak flows, required as input by the erosive sub-model.

The influence of the limited availability of geomorphologic parameters (balanced by the estimation, even reasonable, of some input parameters) as well as of hydrological observations (which even has advised against realistic calibration processes) on the model performance can not be excluded.

However, the availability of proper climatic (allowing set-up of input meteorological data) and GIS sub-routines (helping to process available DEM and themes) together with the user-friendly graphical interfaces in the model software made easy in AnnAGNPS the input data processing. In spite of the large number of input parameters required (more than 100), as for the majority of continuous, physically-based and distributed models, we have remarked a basical easiness of model implementation at the Cannata watershed, thanks to the good availability of geomorphologic and hydrologic information within the experimental data-base as well as the easiness of finding/measuring the majority of input parameters (e.g. meteorological data, soil physical properties). Nevertheless, in some cases processing of simulated hydrologic variables resulted in a time consuming task, especially for surface run-off analysis at event scale.

The model performance could be further improved by optimising algorithms for water balance of soil (in order to improve the simulation of more realistic moisture conditions) or by utilising as input the observed rainfall patterns (at hourly or sub-hourly scales) instead of the synthetic hyetographs utilised at present by AnnAGNPS. Sensitivity analyses, which would allow a more precise estimation of the input parameters to which model response is more sensitive, would be advisable for a better model implementation.

Such improvements, together further research activities aiming at model verification in different environmental conditions, could enhance the model consolidation and stimulate its wider diffusion in professional activities for controlling surface runoff and soil erosion as well as planning mitigation countermeasures.

Author details

Demetrio Antonio Zema[1*], Giuseppe Bombino[1], Pietro Denisi[1], Feliciana Licciardello[2] and Santo Marcello Zimbone[1]

*Address all correspondence to: dzema@unirc.it

1 Mediterranean University of Reggio Calabria, Department of Agro-forest and Enviromental Science and Technology, Italy

2 University of Catania, Department of Agrofood and Environemental System Management, Italy

Notes: The contributions of the authors to this work can be considered equivalent.

References

[1] Aguilar, C., & Polo, M. J. (2005). Análisis de sensibilidad de AnnAGNPS en la dinámica de herbicidas en cuencas de olivar. *In: FJ Samper Calvete y A Paz González, editors. Estudios de la Zona No Saturada del Suelo*, VII, La Coruna, Spain.

[2] AnnAGNPS version 2 user documentation (2001). Available: http://www.ars.usda.gov Accessed 2007 Jan 22.

[3] Arnold, J. G., & Allen, P. M. (1999). Automated methods for estimating baseflow and groundwater recharge from streamflow records. *Journal of the American Water Resources Association*, 35(2), 411-424.

[4] Arnold, J. G., Allen, P. M., Muttiah, R., & Bernhardt, G. (1995). Automated base flow separation and recession analysis techniques, *Ground Water*, 33(6), 1010-1018.

[5] Baginska, B., Milne-Home, W., & Cornish, P. S. (2003). Modelling nutrient transport in Currency Creek, NSW with AnnAGNPS and PEST. Environmental Modelling & Software ., 18, 801-808.

[6] Bagnold, R. A. (1966). An approach to the sediment transport problem from general physics, *Prof. Paper 422-J. U.S. Geol. Surv.*, Reston, VA, USA.

[7] Beven, K. (2006). A manifesto for equifinality thesis. *Journal of Hydrology*, 320, 18-36.

[8] Bingner, R. L., & Theurer, F. D. (2001, 25-29 March). AnnAGNPS: estimating sediment yield by particle size for sheet & rill erosion. Reno, NV, USA. *In: Proceedings of the Sedimentation: Monitoring, Modeling, and Managing, 7th Federal Interagency Sedimentation Conference*, I-1-I-7.

[9] Bingner, R. L., & Theurer, F. D. (2005). *AnnAGNPS technical processes documentation, version 3.2. USDA-ARS*, National Sedimentation Laboratory.

[10] Cronshey, R. G., & Theurer, F. D. (1998). AnnAGNPS: Non-point pollutant loading model. *In: Proceedings of the 1st Federal Interagency Hydrologic Modeling Conference*, 1: 1.9-1.16.

[11] Das, S., Rudra, R. P., Gharabaghi, B., Gebremeskel, S., Goel, P. K., & Dickinson, W. T. (2008). Applicability of AnnAGNPS for Ontario conditions. *Canadian Biosystems Engineering*, 50, 1.1-1.11.

[12] Ganspoel database (2009). Spatially distributed data for erosion model calibration and validation: the Ganspoel and Kinderveld datasets. Available: http://www.kuleuven.be/geography/frg/index.htm. Accessed 2009 Oct 27.

[13] Geter, W. F., & Theurer, F. D. (1998). AnnAGNPS-RUSLE sheet and rill erosion. In: Proceedings from 1st Federal Interagency Hydrologic Modeling Conference Las Vegas, NV, USA.

[14] Indelicato, M., Mazzola, G., Rizzo, N. A., & Zimbone, S. M. (1997). Indagini a scala di bacino su deflussi superficiali ed erosione. In: Proceedings from VI Convegno Nazionale di Ingegneria Agraria Ancona, Italy. , 157-165.

[15] Kliment, Z., Kadlec, J., & Langhammer, J. (2008). Evaluation of suspended load changes using AnnAGNPS and SWAT semi-empirical erosion models. . Catena , 73, 286-299.

[16] Krause, P., Boyle, D. P., & Base, F. (2005). Comparison of different efficiency criteria for hydrological model assessment. *Advances in Geosciences*, 5, 89-97.

[17] Latron, J., Anderton, S., White, S., Llorens, P., & Gallart, F. (2003). Seasonal characteristics of the hydrological response in a Mediterranean mountain research catchment (Vallcebre, Catalan Pyrenees): Field investigations and modelling. *In: Proc. Intl. Symposium: Hydrology of the Mediterranean and Semiarid Regions. IAHS Publ.* [278], 106-110.

[18] Leonard, R. A., Knisel, W. G., & Still, D. A. (1987). GLEAMS: Groundwater loading effects of agricultural management systems. Transactions of ASAE ., 30(5), 1403-1418.

[19] Licciardello, F., Amore, E., Nearing, M. A., & Zimbone, S. M. (2006). Runoff and Erosion Modelling by WEPP in an Experimental Mediterranean Watershed. *In: Owens PN and Collins AJ, editors. Soil Erosion and Sediment Redistribution in River Catchments: Measurement, Modelling and Management. CABI.*

[20] Licciardello, F., Zema, D. A., & Zimbone, S. M. (2009, 17-19 June). Event-scale modelling by WEPP of a Belgian agricultural watershed. Reggio Calabria (Italy). *In: proceedings of XXXIII CIOSTA- CIGR V Conference*, 1741-1745.

[21] Licciardello, F., & Zimbone, S. M. (2002). Runoff and erosion modeling by AGNPS in an experimental Mediterranean watershed. St. Joseph, MI, USA. *In: Proceedings of ASAE Annual International Meeting/CIGR XVth World Congress.*

[22] Loague, K., & Green, R. E. (1991). Statistical and graphical methods for evaluating solute transport models: overview and application. *Journal of Contaminant Hydrology*, 7, 51-73.

[23] Lyndon, M. P., Oduor, P., & Padmanabhan, G. (2010). Estimating sediment, nitrogen, and phosphorous loads from the Pipestem Creek watershed, North Dakota, using AnnAGNPS. *Computers & Geosciences*, 36, 282-291.

[24] Merritt, W. S., Letcher, R. A., & Jakeman, A. J. (2003). A review of erosion and sediment transport models. *Environmental Modelling & Software*, 18, 761-799.

[25] Morgan, R. C. P., Quinton, J. N., Smith, R. E., Govers, G., Poesen, J. W. A., Auerswald, K., Chisci, G., Torri, D., Styczen, M. E., & Folly, A. J. V. (1998). *EUROSEM: documentation and user guide*, Silsoe College, Silsoe, UK.

[26] Nash, J. E., & Sutcliffe, J. V. (1970). River flow forecasting through conceptual models. *Part I. A discussion of principles. Journal of Hydrology*, 10, 282-290.

[27] Nearing, M. A., Jetten, V., Baffaut, C., Cerdan, O., Couturier, A., Hernandez, M., Le Bissonnais, Y., Nichols, M. N., Nunes, J. P., Renschler, C. S., Souchere, V., & Van Oost, K. (2005). Modeling response of soil erosion and runoff to changes in precipitation and cover. *Catena*, 61, 131-154.

[28] Parajuli, P. B., Nelson, N. O., Frees, L. D., & Mankin, K. R. (2009). Comparison of AnnAGNPS and SWAT model simulation results in USDA-CEAP agricultural watersheds in south-central Kansas. *Hydrological Processes*, 23, 748-763.

[29] Polyakov, V., Fares, A., Kubo, D., Jacobi, J., & Smith, C. (2007). Evaluation of a non-point source pollution model, AnnAGNPS in a tropical watershed. *Environmental Modelling & Software*, 22, 1617-1627.

[30] Refsgaard, J. C. (1997). Parameterisation, calibration and validation of distributes hydrological models. Journal of Hydrology ., 198, 69-97.

[31] Renard, K. G., Foster, G. R., Weesies, G. A., Mc Cool, D. K., & Yoder, D. C. (1997). Predicting soil erosion by water: a guide to conservation planning with the revised universal soil loss equation (RUSLE). *Agriculture Handbook* [703].

[32] Sarangi, A., Cox, C. A., & Madramootoo, C. A. (2007). Evaluation of the AnnAGNPS Model for prediction of runoff and sediment yields in St Lucia watersheds. *Biosystems Engineering*, 97, 241-256.

[33] Saxton, K. E., Rawls, W. J., Romberger, J. S., & Papendick, R. I. (1986). Estimating generalized soil-water characteristics from texture. *Soil Science Society of America Journal*, 50, 1031-1036.

[34] Shamshad, A., Leow, C. S., Ramlah, A., Wan Hussin, W. M. A., & Sanusi Mohd, S. A. (2008). Applications of AnnAGNPS model for soil loss estimation and nutrient loading for Malaysian conditions. *International Journal of Applied Earth Observation and Geoinformation*, 10, 239-252.

[35] Shrestha, S., Babel Mukand, S., Das Gupta, A., & Kazama, F. (2006). Evaluation of annualized agricultural nonpoint source model for a watershed in the Siwalik Hills of Nepal. Environmental Modelling & Software ., 21, 961-975.

[36] Steegen, A., Govers, G., Takken, I., Nachtergaele, J., Poesen, J., & Merckx, R. (2001). Landscape and watershed processes. Factors controlling sediment and phosphorus export from two Belgian agricultural watersheds. *Journal of Environmental Quality*, 30, 1249-1258.

[37] Suttles, J. B., Vellidis, G., Bosch, D. D., Lowrance, R., Sheridan, J. M., & Usery, E. L. (2003). Watershed-scale simulation of sediment and nutrient loads in Georgia coastal plain streams using the annualized AGNPS model. *Transactions of the ASAE*, 46(5), 1325-1335.

[38] Taguas, E. V., Ayuso, J. L., Peña, A., Yuan, Y., & Pérez, R. (2009). Evaluating and modelling the hydrological and erosive behaviour of an olive orchard microcatchment under no-tillage with bare soil in Spain. *Earth Surface Processes and Landforms*, 34, 738-751.

[39] Takken, I., Beuselinck, L., Nachtergaele, J., Govers, G., Poesen, J., & Degraer, G. (1999). Spatial evaluation of a physically-based distributed erosion model LISEM. Catena ., 37, 431-447.

[40] Theurer, F., & Clarke, C. D. (1991, 18-21 March). Wash load component for sediment yield modeling. Las Vegas, NV, USA, Paper presented at Subcommittee on Sedimen-

tation of the Interagency Advisory Committee on Water Data. *In: Proceedings of 5th Federal Interagency Sedimentation Conference*, 1:7.1-7.8.

[41] USDA, Soil Conservation Service. (1972). *National Engineering Handbook, Hydrology, Section 4*, 548, Washington DC, USA.

[42] USDA-NRCS (1986). Urban Hydrology for Small Watersheds. United States Department of Agriculture- Natural Resources Conservation Service Conservation Engineering- Division Technical- Release 55.

[43] Van Liew, M. W., & Garbrecht, J. (2003). Hydrologic simulation of the little Washita river esperimental watershed using SWAT. *Journal of the American Water Resources Association*, 39(2), 413-426.

[44] Van Oost, K., Govers, G., Cerdan, O., Thauré, D., Van Rompaey, A., Steegen, A., Nachtergaele, J., Takken, I., & Poesen, J. (2005). Spatially distributed data for erosion model calibration and validation: The Ganspoel and Kinderveld datasets. *Catena*, 61, 105-121.

[45] Wang, S. H., Huggins, D. G., Frees, L., Volkman, C. G., Lim, C. N., Baker, Smith. V., & Denoyelles, F., Jr. (2005). An integrated modeling approach to total watershed management: water quality and watershed assessment of Cheney Reservoir, Kansas, USA. *Water, Air, and Soil Pollution*, 164, 1-19.

[46] Williams, J. R., Jones, C. A., & Dyke, P. T. (1984). A modelling approach to determining the relationship between erosion and soil productivity. Transactions of the ASAE ., 27(1), 129-144.

[47] Willmott, C. J. (1982). Some comments on the evaluation of model performance. Bulletin of American Meteorological Society ., 1309-1313.

[48] Wischmeier, W. H., & Smith, D. D. (1978). Prediction rainfall erosion losses. *USDA Handbook*, 537, Washington D.C.

[49] Young, R. (1994). *AGricultural Non-Point Source Pollution Model, Version 4.03 - AGNPS User's Guide*.

[50] Young, R., Onstad, C. A., Bosch, D. D., & Anderson, W. P. (1987). *AGNPS, Agricultural Non-Point Source Pollution Model. A watershed analysis tool. Conservation Research Report 3*, Washington, D.C., USA, USDA Agricultural Research Service.

[51] Yuan, Y., Bingner, R. L., & Rebich, R. A. (2001). Evaluation of AnnAGNPS on Mississippi Delta MSEA watershed. *Transactions of the ASAE*, 44(5), 1183-1190.

[52] Yuan, Y., Bingner, R. L., Theurer, F., Rebich, R. A., & Moore, P. A. (2005). Phosphorous component in AnnAGNPS. *Transactions of the ASAE*, 48(6), 2145-2154.

Chapter 2

Optimization of Soil Erosion and Flood Control Systems in the Process of Land Consolidation

Miroslav Dumbrovsky and Svatopluk Korsuň

Additional information is available at the end of the chapter

http://dx.doi.org/10.5772/50327

1. Introduction

Extreme hydrological phenomena of recent years have highlighted a well-known fact that it is necessary to pay greater attention to the problems of flood-prevention and soil erosion control on a large part of the Czech Republic. Case study areas are the most endangered territories. The case study area was selected as a case study mainly for its natural conditions and high risk of soil degradation and occurrence of flash floods. Relief, geomorphology, the present state of the complex system of soil properties, the types of agricultural farming practices and land use, are all contributing to accelerated soil erosion and runoff with all its negative impacts on the built-up areas.

The main soil degradation problems in the case study area are soil erosion caused by water, soil compaction and decline in organic matter. Soil erosion is fostered by i) soil degrading (intensive) farming practices such as up and down hill conventional tillage and other conventional agricultural operations on arable land, ii) frequent extreme hydrological events, and iii) a decreasing ability of soils for water retention (decline in organic matter and land conversion). Soil compaction is a problem due to intensive conventional farming on arable land (using heavy machinery). The decline in organic matter results from the constant soil erosion process. Main causes of decline in organic matter are conventional farming practices without using manure and other organic matter. Decline in organic matter causes a decrease of natural crop productivity of soil and decreases yield.

Great runoffs occur on these areas and transform into flood waves in watercourses. Forest grounds are also affected, especially in case of unsuitable transport, wood cut and growth make-up. Solving of the problems of territory protection from unfavourable and damaging effects of overland water flow must therefore begin in catchments areas and particularly during any interference with landscape. Appropriate conservation measures are required to prevent and reduce runoff and soil degradation resulting from intensive agriculture. The adoption of the most appropriate practices and optimisation the farming conservation system it is necessary to carry out analyses and evaluations of the erosion rate and the basic characteristics of runoff in given sub-catchments. This system of evaluation provides information about erosion and runoff risks plots and serves for decision making regarding soil conservation and flood prevention measures. The success of the system of soil conservation depends on suitable technical assistance and support from responsible state organisations (Ministry of Agriculture and Ministry of Environment), sufficient sources of information as well as the ability and willingness of land users to adopt soil conservation measures. The main motivation for farmers to apply soil conservation measures is the economic motivation through financial subsidies along with penalties for farmers if they fail to comply with the rules of the funding program.

Nevertheless, when introducing a soil erosion control and flood prevention measures in a certain watershed, best management practices are mostly to be able decrease of erosion rate but unable to restrict a surface runoff substantially. For that reason it is necessary to apply a whole system of soil conservation measures. In places with long slopes technical and biotechnical soil erosion control practices (primarily of linear character) are necessary. These technical measures are broad base terraces and channels in case study area. These biotechnical measures together with the implementation of grassed courses of concentrated surface runoff (grassed waterways) create an appropriate network of new hydrolines in the watershed. Biotechnical line elements of soil erosion control serve as permanent barriers or obstacles for water runoff and are designed in order to determine, by their location, the ways of land management. Some technical and biotechnical measures could be suitable regarding their technical feasibility, economic efficiency and environmental effectiveness. The spatially and functionally limited soil-conservation system in a given teritorry offers spaces and lines in which it would be possible to locate territorial systems of ecological stability under certain conditions. Soil conservation and flood prevention practices, connected with territorial systems of ecological stability can be characterized as desirable anthropogenic landscape-forming elements. These would form the appearance of the landscape and significantly enhance natural processes in the region. They create suitable biological conditions in spite of the fact that they mostly do not meet qualitative and dimensional characteristics of biocentres and biocorridors.

Highly fragmented land ownership is prevalent in the area Biotechnical and technical soil conservation measures cannot be applied without respecting property rights. Integral parts of any project of soil erosion control (its basic network) are usually line elements for soil erosion control (broad base terraces and channels etc.), which run across individual owners' fields. Therefore it is necessary to identify every owner and discuss with the project and relevant proposals. The greatest interventions with agricultural landscape are land consolidation which, apart from other less important objectives, are designed to completely eliminate or at least partly limit unfavourable effects of runoff (especially soil erosion) and thus to become one of the most important elements of territory organisation and protection. Therefore it was found suitable to design the system of the soil and water conservation in the process of land consolidation in the Czech Republic.

Optimal spatial and functional delimitation of soil erosion control practices in the landscape is one of the basic steps in the plan of comprehenshive land consolidation, in addition to the implementation of a new network of field roads and landscape features enhancing ecological stability. Soil erosion control and flood prevention practices are included in the system of public facilities within the framework of the land consolidation process (where property relations are consistently solved).

The definition of land consolidation is from Act No. 139/2002 Coll., on Reparcelling and Land Authorities and amending Act No. 229/1991 Coll., on the Arrangement of Ownership titles to Land and other Agricultural Assets, as amended. The land consolidation processes in case study area have started in 2005.

This procedure gives solutions to the whole area, both from the aspect of a new land and ownership arrangement (Figure 2) and from the aspect of soil conservation and flood prevention and improvement of environment (Figure 1).

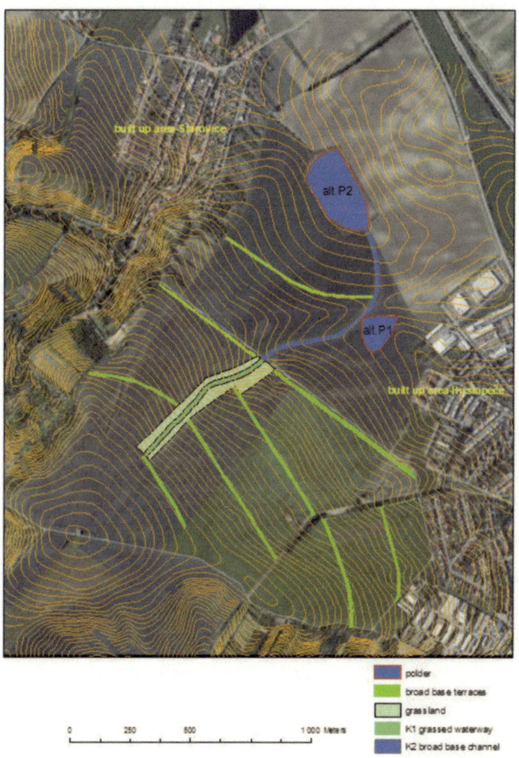

Figure 1. Soil conservation and Flood prevention system.

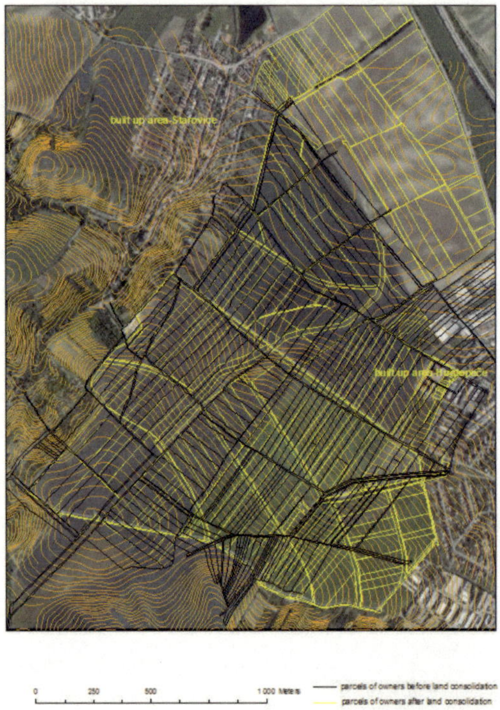

Figure 2. Parcel of owners before and after land consolidation.

Recently, the process of complex land consolidation in the Czech Republic has provided a unique opportunity for improving the quality of the environment and sustainability of crop production through better soil and water conservation. The current process of the land consolidation consists of the rearrangement of plots within a given territory, aimed at establishing the integrated land-use economic units, consistent with the needs of individual land owners and land users.

Integrated territory protection can be reached by controlling runoff by means of design of terraces as a soil erosion control measures. A number of mathematical models, mostly simulation ones, to solve water-management problems have been compiled, some of which include the option of exact mathematical optimization. A certain summary of these models, including their characteristics and application possibilities, were elaborated by Kos (1992). An interesting combination of the application of a simulation and optimization model technique in the elaboration of design of a particular water-management system was described by Major, Lenton et al. (1979), a three-model approach to solve water-management systems was used by Onta, Gupta and Harboe (1991). Benedini (1988) dealt more generally with the design and possible applications of these models. Most likely, an optimization model has not been designed, which would enable to attach territory protection and the measures to eliminate the amount and accumulation of runoff in catchments areas to solving water-management problems.

The created procedure is a universal tool which can be applied for any territory. It enables to find the most suitable combination of all possible alternatives of various erosion controls and flood protection measures under given conditions of each particular site. Such sites do not always have to be ground used for farming. They may also include in forest or urban areas or site arrays in various territories.

2. Method

The optimization process of designing the system of *integrated territory protection* (the *IOU* system) begins with the processing of the system of organisational, agrotechnical, biotechnical and technical measures at individual sites of the case study territory. It is necessary to derive hydrograms of direct runoff from extreme rainfall events for each of these variants. Then it is necessary to elaborate the variants of terraces and other conservation measures on all sections of watercourses and variant of designs of retention protection reservoirs. Not only rivers, streams and brooks are included into the watercourse category within this procedure but also sometimes passed watercourses such as terraces, grass infiltration belts or the lines of stabilisation of concentrated runoff waterways in valley lines.

A selection of the most suitable combination of all prepared variants is listed. With respect to the fact that it is necessary to find optimal dimensions for some of the system elements, there is usually a great number, in case of a continual solving even an infinite number, of possible combinations. It is therefore necessary to use an optimalized mathematical model to find the most suitable combination. This model was created on the basis of a mixed discrete programming (Korsuň et al., 2002, Dumbrovský et al., 2006). Its basic building stones are three generally formulated partial models: *A.* partial model of protective measures at individual sites of the case study area. *B.* partial model of a watercourse. *C.* partial model of a reservoir.

It is possible to shape an *optimization model of integrated territory protection* (*OMIOU*) from these partial models for any particular territory. The partial models are repeatedly inserted into the *OMIOU* as needed so as to exactly copy the modelled system structure. It is necessary to determine in advance one criterion or more simultaneously operating optimization criteria for each optimization function. A whole range of criteria can be determined for a given purpose. These can be taken from the sphere of economy but also from those of ecology, water-management, social etc. However it is necessary to define the most suitable criteria as far as quality is concerned but also to have a chance to quantify the values of each defined criterion. On top of that, it is necessary, in case of several simultaneously operating optimization criteria, to assign each criterion its adequate weight with which it will enter the solving process and which will support its effect on the result, so called a compromise solution in competition with the other criteria.

In creating the procedure of the *IOU* system proposal optimization in connection with the process of territory organisation a requirement of a maximal protection of inhabited and other areas with the exertion of minimal means was formulated for the solving process on the level of land consolidation as one of the suitable optimization criteria. It is a criterion consisting of three simultaneously operating partial economic, but at the same time water-management and socially aimed at their impacts. Criteria include:

- minimization of the average annual damage (material damage: it is estimated that input requirements and conditions will not allow solutions which could lead to losses of human lives) originated by overland runoffs from rainfall events and then by their concentration in watercourses.
- minimization of the average annual economic losses in farming production related to the realisation of proposed protective measures on arable land.
- Minimization of the average annual expenses (the sum of expenses for running and maintenance plus the amortization of the capital goods) of the proposed conservation measures.

Seeing that in most cases they are average annual values, quantified for example in thousands of CZK per year, these criteria can be assigned the same weights 1:1:1 in reflection.

The optimization mathematical model is a system of equations, which model a given system behaviour, the variables in the equation describe a system structure and the dimensions of its individual elements. Non-equations found in each model are transformed into equations by means of additional variables in the course of the model solving process, therefore the term *equation* is used only. The above mentioned partial models were created in the modelling and calculation system *GAMS* (*General Algebraic Modelling System*) in its general form (Charamza, 1993) so it can be used to model any integrated territory protection system. The nature of the solved problems implies that the defining process of all the variables used in the model as positive variables. They can be either continuous ones which are marked x herein after or binary ones (they can take on only 0 or 1 values) marked with the symbol x_B. Other symbols are used to mark variables and coefficients. Activities proceeding in time must be modelled in the whole system according uniform timekeeping.

The partial model A is aimed at terraces and other biotechnical, agrotechnical, and organisation conservation measures in the catchments area of a certain watercourse. These measures are usually designed within land consolidation to decrease overland flow of rainfall events and thus to limit the effects of soil erosion and damage in inhabited territories. The various proposals of protective measures must be elaborated in each individual case before an optimization model is designed (pre-optimization) as pragmatically created systems of various, mutually complementary interventions with the individual catchments area elements. Such a partial catchments area element could be, for example, valley and slope area above one bank of a certain watercourse section in the range from the bank line to the interstream divide line.

The part of runoff from the design rainfall events which will not be caught by the system of catchments area protective measures (*residual runoff*) will concentrate in a particular watercourse and will create a design Q runoff or flood wave. The time T of passage of the design flood wave through a watercourse will be divided into r of equally long *time intervals* (*TI*); time t of the durance of one *TI* will thus be given by the relation $t = T / r$. For the individual *TIs*, partial volumes w_1 of the design flood wave are then quantified, $i = 1, 2,\ldots, r$.

In case of the application of the above mentioned optimization criterion, the following indicators must be quantified for each pre-optimization processed variant of the protective measure set on a partial catchments area element:

- its estimated effect *U* expressed financially as an average annual level of damage on land, growth, buildings, roads etc. which will occur after the variant has been realised (*residual damage*),
- estimated average annual economical loss *E* in farming production related to the realisation of the proposed measures on arable land.
- realisation costs of a particular variant and its average annual own costs *N*,
- the amount of residual runoff O_i into a watercourse in the individual *TIs*.

These data represent input information for the partial model *A*. In the course of the optimization process, only one – optimal – variant with the most suitable indicators will be chosen from thus prepared variants of systems of protective measures for each partial catchments area element. Residual runoffs concentrating in a watercourse runway from the watercourse adjacent partial areas protected by optimal systems of measures will cause a gradual accretion of a flood wave passing through the watercourse. The protection from damage which could be caused by this flood wave will be provided by the protective measures on the watercourse and retention protective reservoirs as mentioned later (the partial models *B* and *C*).

Binary variables can be used for modelling of individual variants of protective measure systems in each of the partial catchments area elements in a discrete way. The total number of catchments area elements will be *m*. If, for example, *n* variants of protective measure systems of a dth catchments area element are modelled by relations to binary variables $x_{B1dp} \in \{0, 1\}$, $d = 1, 2, \ldots, m$, $p = 1, 2, \ldots, n$, the effects of these measure systems for this catchments area element can be write into the model using the following equations:

the equation of protective effects (residual damage)

$$x_{Ud} = \sum_{p} U_{dp} \cdot x_{B1dp} \tag{1}$$

the equation of economic damage

$$x_{Ed} = \sum_{p} E_{dp} \cdot x_{B1dp} \tag{2}$$

the equation of own costs

$$x_{Nd} = \sum_{p} N_{dp} \cdot x_{B1dp} \tag{3}$$

the equation of residual runoff, i.e. contribution of a d th catchments area element to the flood wave volume on a particular watercourse section in i th TI

$$x_{Oid} = \sum_{p} O_{idp} \cdot x_{B1dp} \tag{4}$$

for $i = 1, 2, \ldots, r$,

$d = 1, 2, \ldots, m$,

$p = 1, 2, \ldots, n$,

where x_{Ud} is the total residual damage in a d th catchments area element,

$x_{E,Ud}$ is the total economic loss in a d th catchments area element,

x_{Oid} is the total residual runoff from a d th catchments area element in i th TI.

Because only one of the protective measure system variants can enter the solving process, the following condition must be valid for the sum of all the binary variables of a d th catchments area element:

$$\sum_{p} x_{B1dp} = 1 \tag{5}$$

The partial model B captures the passage of the design flood wave through the watercourse sections. The sections are either left in their present state, the optimization of a river bed or a contour furrow systems design (including the building of protective dams), or the reconstruction an earlier carried out adjustment or protective dams may be required. A watercourse section can also be a water or dry protective reservoir which will be modelled in a way mentioned in the partial model C description.

Flood damage that can occur is quantified for each watercourse section during its modelling. Further, runoffs from the section are calculated in the individual *TIs* of a flood wave passage. With respect to the overland flow from the initial section profile to the last one, it is necessary to determine a time shift which will affect collisions of flood waves on the main watercourse and at the mouths of its tributaries. The mean value of the runoff volume which can be found in a section (in a river bed or also in an inundation territory) in the course of i th TI is at the position of the basic section variable. The values of the other variables are related to this variable: the variables of the water flowing through the section, time of concentration, the level of flood damage in the section, and the level of runoff from the section. The courses of these non-linear functions are derived from the watercourse pre-optimization variant designs. They are replaced with linear function part by part in the optimization model. The formulation of particular equations is mentioned in Chapter 5.1 of Patera, Korsuň et al. (2002).

The partial model C is outlined for a designed multipurpose water reservoir with unknown capacities of spaces protective controllable x_{OO}, protective non-controllable x_{ON} and total x_V. The necessary volumes of the spaces of dead storage $S \geq 0$ and active storage capacity $Z \geq 0$ are constant – these values result from other than protective requirements. The objective of analysis is to find its dimension which, respecting the requirement to create the spaces S and Z, with its protective spaces will ensure the reduction of culminated runoff from the reservoir to its optimal level during the passage of the design flood wave. In cases when the designed water reservoir has only a protective function, the value of the Z variable is zero; the values of both variables are zero $S = Z = 0$ for a dry protective reservoir.

The unknown volume of the total reservoir space is a variable, whose value which is limited from above by the maximal value V_{max} corresponding with the biggest realisable variant of the reservoir design during the pre-optimization solutions. From below it is limited by the minimal variant, still acceptable for practice, with the total volume V_{min}.

We cannot forget a situation when building a reservoir will not be acceptable due to the used optimization criteria. It is therefore necessary to introduce a binary variable $x_{B2} \in \{0, 1\}$ into the set of variable values. If this variable has a zero value, the reservoir will not enter the solving process, if $x_{B2} = 1$, the entry of the reservoir into solving is cleared. Then the volume of the total reservoir space (without evaporation and percolation) must correspond with the following conditions

$$x_V = (S + Z) \cdot x_{B2} + x_{OO} + x_{ON} \tag{6}$$

$$V_{min} \cdot x_{B2} \leq x_V \leq V_m \tag{7}$$

The equations modelling the passage of the design flood wave through a dam profile, the calculations of the volumes of individual reservoir spaces and of necessary financial means are described in Chapter 5.1 of Patera, Korsuň et al. (2002). The partial model C can be also used for already an existing reservoir with a constant volume of the total space.

The model compilation from the fore mentioned partial elements in the presented form requires the introduction of a set of concrete coefficients and variables into the model for the model equation system to copy completely a particular system of *IOU*. These coefficients and variables should be derived from the pre-optimization processed background materials. In the case of non-standard requirements of an *IOU* system structure, it is necessary to introduce other equations to the model. Such new equations would capture these requirements. The model solving process in carried out on a computer by means of some of the *GAMS* system tools.

3. Material

To verify the function and potential of the already described optimization procedure, a system of integrated territory protection was chosen that was proposed within the framework of land consolidation on the case study area between the town of Hustopeče and the village of Starovice in the Czech Republic (see Figure 1). The declining ground in this region is mostly used as arable land. Overland flow is concentrated into its main waterway, which enters the residential parts of Hustopeče. Considerable, and frequently repeating damage, is caused by soil erosion on farm crops, sediment transport from arable land and especially by flooding parts of the town.

The proposed system of integrated protection of this farming territory and town is based on a system of technical-biotechnical, organisational and agrotechnical soil erosion control measures on arable land and of two conservation measures: 1. transfer of concentrated runoff from the drainage furrow or channel *K1* in the main valley line over the terrain into the adjacent valley line and creating a channel *K2* entering watercourse, 2. building a dry protective reservoir (polder) *P1* to catch parts of runoffs from the main valley line and another polder on the channel *K2* in the adjacent valley line above the village of Starovice.

The *IOU* system design for the given territory is based on the situation which would occur during a rainstorm with hundred-year periodicity (design rainfall). The protective measures with pre-optimization were designed in ten different variants, volume and cost (own costs) functions were derived for both the polders. It is estimated that *P1* polder filling, which is a side basin for the channel *K1*, will proceed through the channel side overfall. For the individual soil erosion control measure alternatives volumes and accumulation of overland runoffs, derived from the design rainfall, in the form of runoff hydrograms from two catchments areas: from the polder *P1* catchments and from the polder *P2* catchments. The passage of runoff waves through dam profiles of both polders takes from 510 minutes in the alternatives 1 and 2 to 195 minutes in the alternatives 9 and 10. It requires limiting the culmination water passages in the river beds below the two polders: below the *P1* this passage (runoff from the *P1*), which will enter the city sewerage system in Hustopeče, should not exceed 0.125 $m^3.s^{-1}$, below the *P2* the passage limit should be, with regard to the protection of Starovice, chosen at 1.5 $m^3.s^{-1}$ at most and in variants of 1.0 and 0.5 $m^3.s^{-1}$ to determine the effect of this passage size on the *IOU* system optimal solution.

The optimization model consists of 3,506 equations with the total of 1,673 structural variables, 539 of which are binary variables. The model objective function (optimization criterion) minimises the sum of average annual values of flood damages, economic losses and biotechnical measures and polders own expenses in the proportion of 1:1:1. It is ensured that only one protection system alternative can enter the *OMIOU* optimal solution in both the catchments areas, but it can be different for each of the catchments. These alternatives are marked as *A1* and a particular alternative number for the *P1* polder catchments, the *P2* polder was allocated symbol 2 in a similar way. The polders can enter the solution but they also do not have to. The runoff wave from the *P1* polder catchments may be partly or completely transferred into the *P2* polder. Permissible maximum of water depth in the *P1* polder is 5.0 m, it is 4.34 m in the *P2* polder. The *P1* polder low outlet dimensions (the inside diameter of outlet pipeline of a round shape) d = 200 mm, there is a possibility of choice from d = 200, 300, 400, 500 or 600 mm for the *P2* polder.

4. Results and discussion

The model function and behaviour were examined first in relation to the project research objectives. Then the possibilities of experimentation on the model of the designed system were tested (Korsuň et al., 2002). The optimization process was carried out with the three above listed values of admissible maximal runoff from the *P2* polder and then in an experimental way with various runoffs from both catchments areas: with real runoffs derived from hundred-year rain storms for the individual variants of conservation measures in both the catchments areas, and with fictive multiples of these runoffs.. Variants with other changes in input conditions (e.g. without the polders entering the solving process) were calculated for the same reason. The results of these solutions are not listed here. Optimal solutions of variants No. 1, 3 and 5 correspond with the real state of the input conditions. These solution results were derived from overland flows from the grounds and from three real values of admissible maximal runoff from the *P2* polder above Starovice. The results of following experiments on the optimization model have led to a number of interesting findings. However, the most important finding is the fact that the experimental locality can be protected as required without any interference into plant production conditions, i.e. without any (on site) economic loss on the produce only by conservation measures themselves: by draining overland and hypodermic runoffs through contour furrows and channels in the *P2* polder. This protective system design is valid only provided the applied optimization criterion is kept. The resulting design can be different in the case of any change to the criterion (e.g. the changes in the weights of the three used partial criteria) or in case of the application of a different criterion.

5. Conclusion

The results of the practical application of the optimization procedure in designing terraces and retention reservoirs within integrated territory protection verify its functionality and applicability. In cases when it is not clear in advance which of the potential torrential rainfall could be the most dangerous, the model will provide solutions with all chosen rainfalls types for the result to comply with the territory protection requirements.

The created model can be used to find either one optimal solution or, in case it is necessary to verify the position of the optimal solutions with the changes of some input conditions and requirements, more times in more versions with variables and coefficients modified by these changes. The possibility of multiple application of this model and to obtain a whole set of optimal solutions visualises much better the character and behaviour of the designed system in reactions to modifications of the input conditions and requirements and thus enables to improve significantly the process of making decisions about the design final shape.

A great advantage of the model lies in the general formulation of its components – partial models of conservation measures at individual sites of the experimental locality, watercourse and reservoir. This should enable its problem-free application for optimization design of integrated territory protection under any conditions and at any site

Author details

Miroslav Dumbrovsky* and Svatopluk Korsuň

*Address all correspondence to: dumbrovsky.m@fce.vutbr.cz

Brno University of Technology, Faculty of Civil Engineering, Department of Landscape Water Management, Czech Republic

References

[1] Benedini, M. (1988). Developments and possibilities of optimisation models. *Agric. Water Manag.*, 13, 329-358.

[2] Dumbrovský, M., et al. (2003). Optimisation of the system of soil and water conservation for runoff minimizing in certain watershed in the process of land consolidation (in Czech). *Final research report, NAZV-QC1292*, VÚMOP, Praha.

[3] Charamza, P., et al. (1993). Modelling system GAMS (in Czech). MFF UK, Praha.

[4] Korsuň, S., et al. (2002). Creating and verification of the model for optimisation of soil and water conservation (in Czech). *Final research report, A01-NAZV-QC1292*, FAST VUT, Brno.

[5] Kos, Z. (1992). Water management systems and their mathematical models during climate changes (in Czech). *Vod. Hosp.*, 7, 211-216.

[6] Major, D. C., Lenton, R. L., et al. (1979). Applied water resource systems planning. . Prentice- Hall Inter., Inc., London.

[7] Onta, P. R., Gupta, A. D., & Harboe, R. (1991). Multistep planning model for conjunctive use of surface and groundwater resources. *Jour. Water Res. Plan. Manag.*, 6, 662-678.

[8] Patera, A., Váška, J., Zezulák, J., Eliáš, V., Korsuň, S., et al. (2002). Floods: prognosis, water streams and landscape (in Czech). *ČVUT / ČVVS*, Praha.

Chapter 3

Terrain Analysis for Locating Erosion Channels: Assessing LiDAR Data and Flow Direction Algorithm

Adam Pike, Tom Mueller, Eduardo Rienzi,
Surendran Neelakantan, Blazan Mijatovic,
Tasos Karathanasis and Marcos Rodrigues

Additional information is available at the end of the chapter

http://dx.doi.org/10.5772/51526

1. Introduction

Terrain analysis can be used to locate concentrated flow erosion (e.g., ephemeral gully erosion) across landscapes. For example, studies have found that ephemeral gullies were likely to occur when field specific thresholds were exceeded for the following terrain attributes: the product of upslope area, slope, and plan curvatures [1]; topographic wetness index, upslope area, and slope [2]; and the topographic wetness index and the product of the upslope area and slope [3]. Another study used a cartographic classification and threshold procedure for erosion channel identification [4].

An alternative approach utilized logistic regression and artificial neural network procedures to predict where erosion channels would appear in agricultural fields based on digital terrain attributes [5]. With leave-one-field-out validation, it was determined that the more simple logistic regression was more appropriate because it performed as well as the non-linear neural network procedure. In a follow up study, erosion channels predicted from terrain attributes derived from 10-m US Geological Survey (USGS) digital elevation models (DEMs) were compared to those derived from DEMs created with survey-grade real-time kinematic (RTK) Global Positioning System (GPS) data [6]. The USGS models identified most eroded features but the RTK analyses delineated them more clearly. The authors concluded that the USGS predictions were adequate for many agricultural applications because creating DEMs with RTK was relatively costly while USGS data was freely available on the Internet for most of the United States. A graphical representation illustrates how the logistic regression analysis [5,6] can be fit (step 1) and then applied (step 2) in Figure 1.

In another study, the model from reference [5] predicted where eroded waterways would occur 223 km away in a different physiograpic region of Kentucky where there were marked textural and parent material differences between the soils [7]. The locations of the Outer Bluegrass [6,7] and Western Coal Fields studies are shown in Figure 2.

1.1. RTK GPS, LiDAR, and USGS DEM Accuracy

The performance of terrain analysis applications depends on the source and quality of the elevation information. Very accurate elevation measurements can be obtained with survey-grade RTK GPS (e.g., horizontal rmse < 2.2 cm).

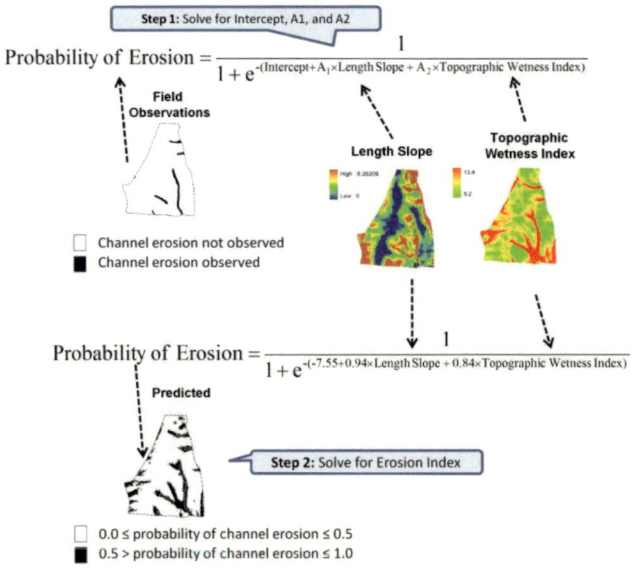

Figure 1. Description of the basic two-step logistic regression analysis procedure used in references [5,6]. The first step involves fitting a model with field observations of soil erosion as a function of terrain attributes. The second step involves applying the model. In this example, the model was applied to the same field as it was fit. However, references [5,6] used a leave one-field out validation and fit the model to four fields and applied it in a fifth. See equation 3 in methods section for a mathematical definition of the Length-Slope terrain attribute.

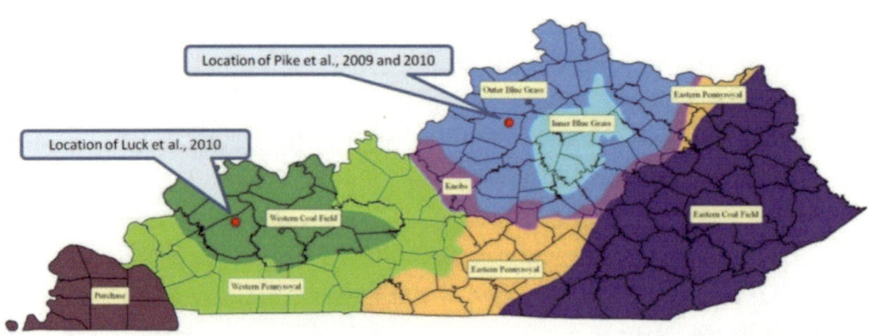

Figure 2. Locations of the Pike et al. (2009 and 2010) [5,6] and Luck et al. (2010) [7] studies. The Kentucky Physiographic regions are labelled and colored.

This usually involves setting up one RTK receiver as a base station, to broadcast via radio corrections to a mobile receiver, which calculates elevation on-the-fly. Elevation data can also be obtained with light detecting and ranging (LiDAR) systems mounted at the bottom of airplanes with a pulsing laser scanning rapidly from side to side. Detectors determine the time required for laser pulses to bounce back in order to calculate the distance between the aircraft and the ground surface. With simultaneous in-flight RTK GPS measurements, estimates of ground elevation are obtained at a high spatial intensity (e.g., several points per square meter). In one study [8], RMSEs for LIDAR were determined to be 33.3 cm in short grass. The USGS National Elevation Dataset (NED) includes Level-2 DEMs for most of the United States, which in many cases were reinterpolated from USGS topographic contours. These contour maps were originally created by the USGS using stereo orthophotogrammetry. Level-2 DEMs have an accuracy of one-half the contour width (e.g., the contour width was for the studies in [5,6] was 304 cm).

Of the three elevation datasources discussed, RTK GPS is one of the most accurate methods for creating ground surveys. Unfortunately, these systems are relatively expensive and obtaining ground measurements and data processing may be labor intensive. USGS DEMs are an attractive option because they are free; however, they do not identify erosion pathways as clearly as the RTK data [6]. Elevation data obtained with LiDAR is relatively inexpensive on a per area basis and are currently being obtained by various government agencies not necessarily associated with agriculture, but urban planning, transportation, and geophysics applications. LiDAR has a great potential to aid in the identification of concentrated flow pathways [9,10].

1.2. Flow Direction Algorithm

Terrain attributes (e.g., length–slope, topographic wetness index) require estimates of the upslope contributing area for each cell in the DEM. This necessitates the calculation of single or multiple flow direction for each cell in the DEM. The most common single direction flow model is the deterministic eight-neighbor (D8) procedure. Multiple direction flow models include fractional deterministic eight-neighbor (FD8), digital elevation model networks (DEMON), and deterministic infinity (D∞). There are a number of software programs that can calculate terrain attributes, including ArcGIS, Grid-Based Terrain Analysis Programs for the Environmental Sciences (TAPES-G), and Terrain Analysis Using Digital Elevation Models (TauDEM). ArcGIS, TauDEM, and TAPES-G predict D8 single direction flow. The TAPES-G program can also estimate multidirectional flow with FD8 and DEMON while TauDEM program can estimate flow with D∞.

O'Callaghan and Mark [11] developed the D8 model that routs flow from each cell to one of its single eight neighbors in the cardinal or diagonal direction with the steepest grade. However, the D8 method does not accurately model divergent flow in ridge areas and it produces unrealistic parallel flow lines [12]. The FD8 multidirection method [13] allows flow to be routed to more than one of its eight neighbors in an amount proportional to the slope gradient between the center cell and the adjacent eight neighbors. Once the upslope contributing area for each cell exceeds a threshold (i.e., maximum cross grading area), the FD8 procedure switches to D8 flow, allowing the modeling of both divergent and convergent flow [12]. Lea [14] developed an aspect driven kinematic routing algorithm that was no longer restricted to the eight nearest neighbors (i.e., cardinal and diagonal directions). This algorithm inspired both the D∞ and DEMON methods. The DEMON stream tube method developed by Costa-Cabral and Burges [15] is computationally intensive and involves routing flow downstream along tubes that expand and contract in a way where the tubes do not necessarily coincide with the cell boundaries [12]. The D∞ algorithm first calculates flow direction from the infinite set of possible flow directions around each cell, not just in the 8 cardinal directions.

The terrain attributes in references [5-7] were calculated with TAPES-G utilizing the FD8 method. TAPES-G still exists but it is no longer being supported and updated by the developers. The Windows installation only works on legacy versions of ArcGIS. The preferred approach for terrain analysis today has become TauDEM with the D∞ flow direction algorithm.

1.3. Concentrated Flow Erosion, Grassed Waterways and The Environment

Agriculture contributes 80% of the excessive phosphorus (P) flowing to the Gulf of Mexico from the Mississippi River Basin. Among the states contributing to this environmental impact, Kentucky is the sixth largest contributor of nitrogen and phosphorus to the Mississippi River Basin [16]. Sediment is the leading cause of impairment of Kentucky's rivers and streams, impacting over 4,329 linear km with agriculture being the primary contributing source [17]. The problem of sedimentation, often underestimated, is enormous. It not only adversely impacts aquatic life [18] and impaires ecological and economic functioning of drainage systems, wetlands, streams, and rivers but also increases the risk of flooding and the need for water treatment. Properly constructed grassed waterways (NRCS Code 412), with appropriately sized vegetative filters on either side reduce considerable runoff, nutrient, and sediment delivery to surface waters. Specifically, they reduce ephemeral gully erosion, which is a substantial but often overlooked problem that accounts for about 40% of all erosion in agricultural fields [19]. In one study, the installation of grassed waterways led to reductions of 39 and 82% in runoff and sediment, respectively [20]. In another study, grassed waterways reduced dissolved reactive phosphorus losses by factors ranging between 4 to 7 and particulate phosphorus by factors ranging between 4 and 10 [21]. Because grassed waterways are so effective, the USDA Conservation Reserve Program (CRP) and Environmental Quality Incentives Program (EQIP) provide funding for this and other conservation practices.

1.4. Research Objective

The objective of this study was to compare how predictions of concentrated flow erosion performed with LiDAR, survey grade RTK GPS, and 10-m USGS DEMs, and using the D8, FD8, DEMON, and D∞ flow direction models.

2. Materials and Methods

This study was conducted in five fields in Shelby County located in the Outer Bluegrass physiographic region of Kentucky. Field A (38°17′ N, 85°9′ W), B (38°18′ N, 85°11′ W), C (38°20′ N, 85°12′ W), D (38°20′ N, 85°11′ W), and Field E (38°20′ N, 85°14′ W), were 23, 36, 11, 57, and 33 ha in size, respectively. The fields had been in a no-till, corn (Zea mays L.), wheat (Triticum aestivum L.), and double-crop soybean [Glycine max (L.) Merr.] or corn-wheat rotation for more than 20 yr. Soils in this region developed primarily from limestone residuum overlain with pedisediment from limestone weathered materials and loess [22].

2.1. Field observations

One of the co-owners of the Worth and Dee Ellis Farms was trained by the NRCS to identify eroded zones from concentrated water flow. The farmer delineated and mapped the eroded ephemeral gullies in the five study fields using a DGPS system. Subsequently, the farmer installed grassed waterways in these areas but did not reshape them as is typically the case when installing these conservation structures. This is an important distinction because reshaping these locations would have changed the terrain attributes, making the analyses presented in this chapter difficult or impossible to interpret.

Because these field observations were conducted by the farmer, we asked the NRCS district conservationist to validate the erosion delineations in the field. The conservationist visited the fields in 2008 and examined 42% of the eroded channels delineated by the farmer in the five study fields. He determined that all the channels would have been eligible for cost support for grassed waterways through the continuous USDA Conservation Reserve Program CRP program if they did not already have waterways installed.

2.2. Acquisition of Elevation Data, DEM Creation, and Terrain Analysis

Survey grade RTK GPS equipment was used to collect elevation data from Field D in 2000 [23], Fields E and B in 2004 [24], and in Fields A and C in 2007 [5]. Dual frequency Trimble AgGPS 214 (base station) and Trimble 5800 (rover) RTK receivers were used to create surveys in Fields A, B, C, and E. The survey for Field D was created with two single-frequency Trimble 4600 receivers and elevation was determined during post processing. The surveys were created at varying intensities with elevation measurements spaced approximately every 3 (Field D) and 4 m (Fields A, B, C, and E) along the direction of travel during the survey. There were 7.5 (Field D) and 12 m (Fields A, B, C, and E) between survey passes. According to the Trimble data specifications, vertical errors for all receivers were expected to be <2.2 cm because the baselines were <1200m [25,26]. The LiDAR survey was created (courtesy of Photo Science) in November of 2009 after harvest with soybean residue and stubble on the ground. The airplane height above ground was 1.219 km and speed was 185 km hr^{-1}. The estimated horizontal and vertical accuracies were 16 and 11 cm, respectively. The average elevation point density was 2.82 points m^{-2}. PhotoScience converted the raw LiDAR data into a triangulated irregular network (TIN). Then they converted the TIN into a 1-m raster file using natural neighbor interpolation. After the 1-m LiDAR DEM was delivered to the University of Kentucky from PhotoScience, the grid was sampled to a 4 by 4 meters. Elevation data on 9.1-m grids for these fields were obtained from Kentucky Division of Geographic Information (KDGI) (http://technology.ky.gov/gis/). The USGS DEMs had been previously re-interpolated by KDGI from 10-m to 9.1 meters so that they could distribute in the Kentucky State Plane coordinate system. The 9.1-m USGS raster was converted to a point file format. The raw RTK GPS and USGS points data were converted to 4 by 4-m grids with the ArcGIS (ESRI, Redlands, CA) TOPOTORASTER command (drainage enforcement option was not used for the reasons described in reference [5]). To smooth the data, 1-m contour maps were created from the 4 by 4-m RTK, LiDAR, and USGS DEMs with the ArcGIS spatial analyst extension. Then the TOPOTORASTER command was used to convert the contours back into new 4-m DEMs.

Next, TAPESG was used to remove sinks (depressions) and calculate terrain attributes for the RTK and USGS datasets using the FD8 and DEMON flow direction algorithms. TauDEM was used to remove pits and calculate terrain attributes for the RTK, LiDAR, and USGS datasets using the D8 and D∞ flow direction algorithms. Primary terrain attributes included slope (β), plan curvature, upslope contributing area, and specific catchment area. Slope was calculated using the finite distance algorithm in TAPES and the D∞ algorithm (i.e., slope could be in any direction, not just in the cardinal and diagonal directions) with TauDEM. For the FD8 method, the maximum cross-grading area [12], was hardcoded to a value of 50,000 m^2 by the authors of the TAPESG for Windows software. Each cell in the DEM had a flow width dependent on the direction of flow entering that cell from the eight neighbors. For the TauDEM D8 and D∞ procedures, flow width was assumed to be 1 grid increment in all directions. The flow width for FD8 and DEMON algorithms is described in detail in reference [12]. The specific catchment area adjusts the upslope contributing area to account for flow width and was calculated as follows:

$$\text{Specific Catchment Area}=\frac{\text{Upslope Contributing Area}}{\text{Flow Width}} \tag{1}$$

Secondary terrain attributes (i.e., those computed from two or more primary attributes) were also determined with TAPESG. This included the topographic wetness index calculated as

$$\text{Topographic wetness index}=\ln\left(\frac{\text{Specific catchment area}}{\tan\beta}\right) \tag{2}$$

and the length-slope factor calculated as

$$\text{Length-Slope}=1.4\left(\frac{\text{Specific Catchment}}{22.13}\right)^{0.4}\left(\frac{\sin\beta}{0.0896}\right)^{1.3}. \tag{3}$$

The length-slope terrain attribute (Eq. [3]) was developed to be an estimate of the length-slope factor from the Revised Universal Soil Loss Equation (RUSLE), and index of potential erosion [27]. Unfortunately, this index does not estimate the true length-slope well on longer and steeper slopes [22] but still has adequate utility.

2.3. Statistical analyses

A new variable, "ErWater," was created that was assigned a value of 0 if the grid point observations were not from areas with evidence of concentrated flow and a value of 1 if there was evidence. The terrain attributes for each 4 x 4 grid from all five fields (n= 99,505) were exported from ArcGIS in a database format so they could be read by SAS. The sample module in SAS Enterprise Miner 4.3 were used to random subsample the dataset with an equal number of observations from each field and areas within the fields with and without concentrated flow erosion. So after subsampling, the final dataset included 1450 observations (145 from each of the eroded areas and non eroded areas in each of the five fields studied). SAS PROC LOGISTIC was used for logistic regression with the 1450 point subset used as the input dataset. The SCORE option was used to obtain full dataset predictions with all 99,505 records. The topographic wetness index, the estimated length-slope for the Universal Soil Loss Equation, and plan curvature were included as predictor variables. The dependent variable was the newly created ErWater variable. Proc FREQUENCY was used to create confusion tables for the score datasets and the results were reported in percentages. The confusion tables presented in this chapter can be interpreted according to the guide presented in Table 1.

Actual Field Status	Predicted	
	NE (non-eroded)	**E (eroded)**
NE (non-eroded)	Correctly classified non-eroded areas	Incorrectly classified un-eroded areas (Type 1 errors)
E (eroded)	Incorrectly classified eroded Areas (Type II errors)	Correctly classified eroded areas

Table 1. Confusion table interpretation guide.

The scored logit data (i.e., 0's and 1's) were then exported from SAS into ArcGIS where the point data were converted to raster format for display in GIS.

3. Results and Discussion

3.1. Performance of LiDAR data with D∞ and D8 Flow Direction Models

Logistic regression models and statistical tests are given in Table 2. The topographic wetness index, length-slope, and plan curvature values were all highly significant in the models. The discretized output of the LiDAR D8 (Figure 3) and D∞ (Figure 4) models indicated a high correspondence between the eroded waterway boundaries and the black shaded areas (i.e., areas with probability of concentrated flow erosion > 0.5 and ≤ 1.0). The average of the combined type 1 and type 2 error rates across the five fields was 8% for D8 and 9% for D∞ (Table 2). This data can be interpreted using Table 1 provided in the methods section.

The LiDAR predictions with D8 and D∞ were excellent (Figure 3 and 4). Interestingly, the D8 approach differentiated between the eroded and non eroded areas in the lower central portion of Field A (Figure 3). In this area, an eroded feature that appears to be an island can be observed by examining the polygon boundaries. The water flowing across this island area disappeared underground and reemerged in the waterway below. The terrain modeling predicted high upslope contributing area and therefore high topographic wetness and length-slope index indices below this island area; however, plan curvature values did not have large negative values (indicating concavity) because there was no erosion area as determined during field observations with the NRCS conservationist. The D8 model differentiated this area because most of the prediction weight from this model came from plan curvature values. This was apparent from the Wald Chi Square statistic which was much greater for the D8 (180) than D∞ (71) analyses.

Flow Direction Method	Variable	Parameter Estimate	Wald Chi Square	
D8	Intercept	-3.58	68	**
	Topographic Wetness Index	0.303	24	**
	Length-Slope	0.758	45	**
	Plan Curvature	-6.81	180	**
D∞	Intercept	-5.97	111	**
	Topographic Wetness Index	0.573	59	**
	Length-Slope	1.06	75	**
	Plan Curvature	-4.75	71	**

Table 2. Logistic regression model parameters and tests for the LiDAR data analyses for the TauDEM D8 and D∞ flow direction models.

3.2. Impact of data cleaning

We considered what would have been the result if we had not contoured and rasterized the data. The average combined type-1 and type-2 error rate for D8 and D∞ was 8% for the smoothed data (Table 2) and 12% for the unsmoothed data. The logistic regression analyses did not differentiate as clearly eroded and non-eroded areas when the data were not smoothed. This is very apparent when comparing the unsmoothed analyses of Fields D and E (Figure 5) with smoothed analyses (Figure 3).

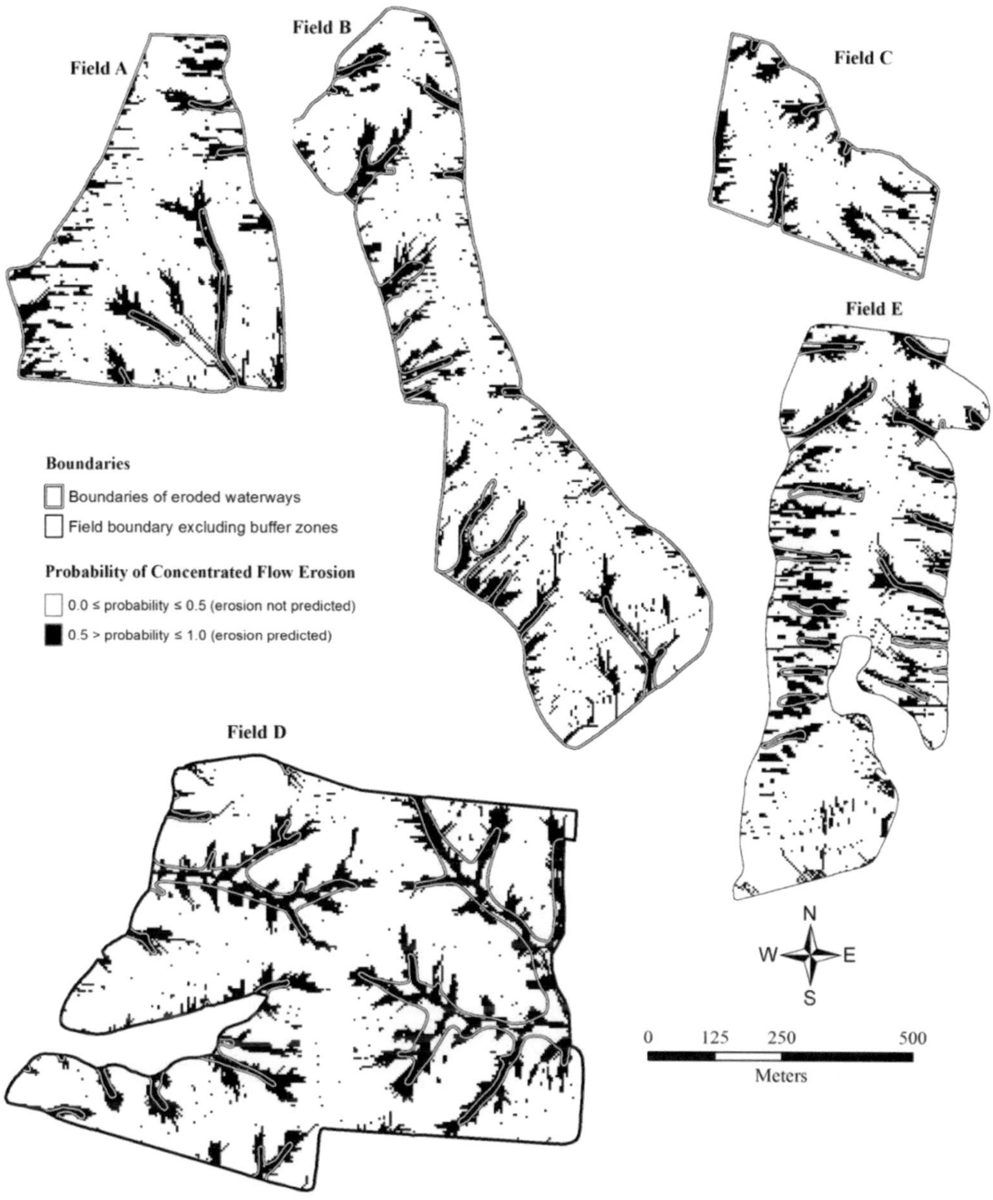

Figure 3. Discretized erosion probability maps derived from LiDAR measurements overlain by the boundaries of the observed concentrated flow pathways. Terrain attributes were calculated with TauDEM using the D 8 flow direction algorithm.

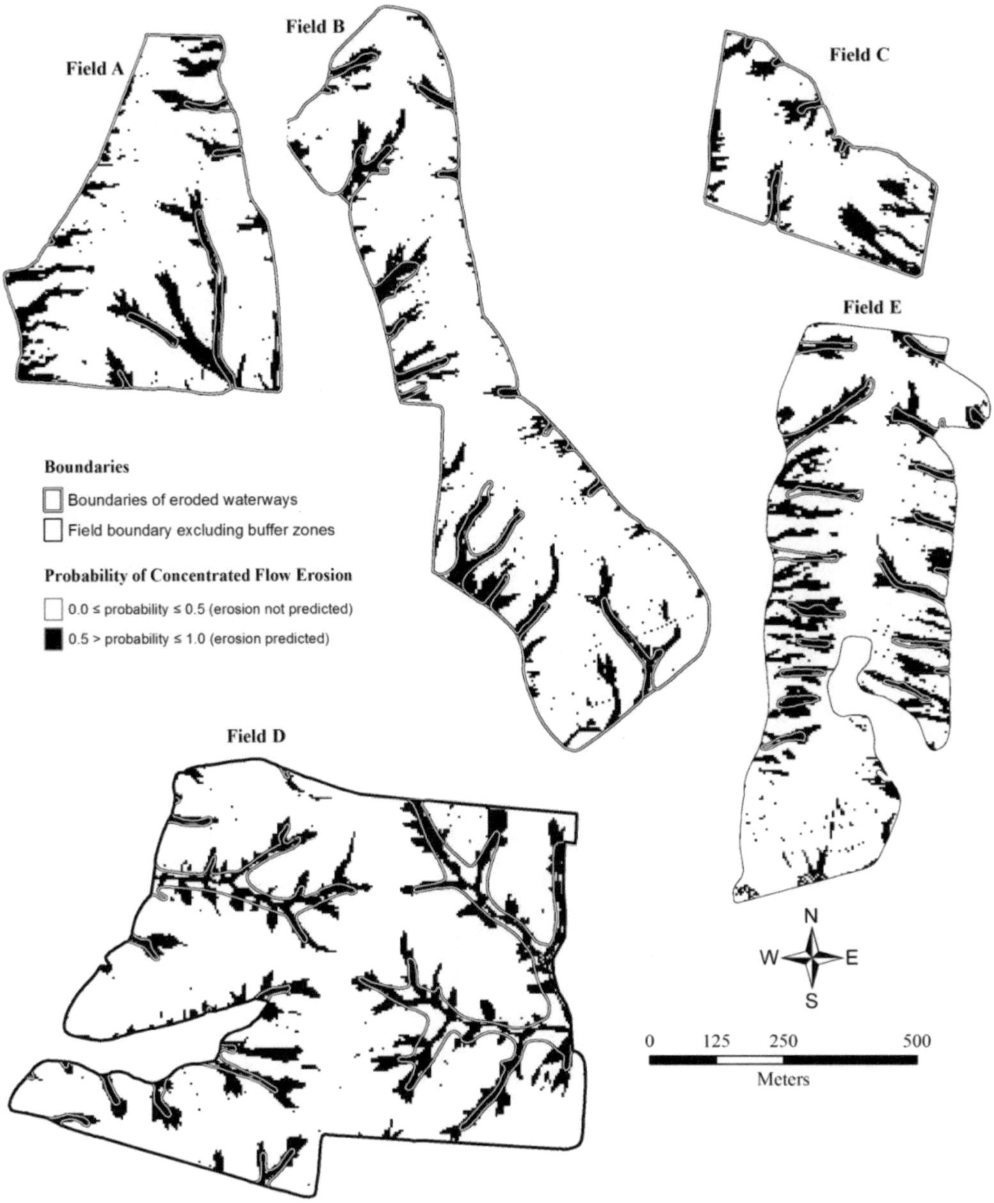

Figure 4. Discretized erosion probability maps derived from LiDAR measurements overlain by the boundaries of the observed concentrated flow pathways. Terrain attributes were calculated with TauDEM using the D ∞ flow direction algorithm.

http://dx.doi.org/10.5772/51526

Flow Direction Algorithm		D8		FD8	
		Predicted			
Field	**Actual Field Status**	**NE**	**E**	**Ne**	**E**
A	NE	81	16	80	17
	E	1	3	1	3
B	NE	83	11	81	13
	E	1	5	2	5
C	NE	79	18	79	18
	E	0	2	0	2
D	NE	78	10	76	13
	E	4	7	4	8
F	NE	75	17	73	18
	E	3	6	3	6

Table 3. Confusion table for the TauDEM D8 and D∞ flow direction models. The values are given in percentages of the total observations from each field. This table can be interpreted using the guide presented in Table 1.

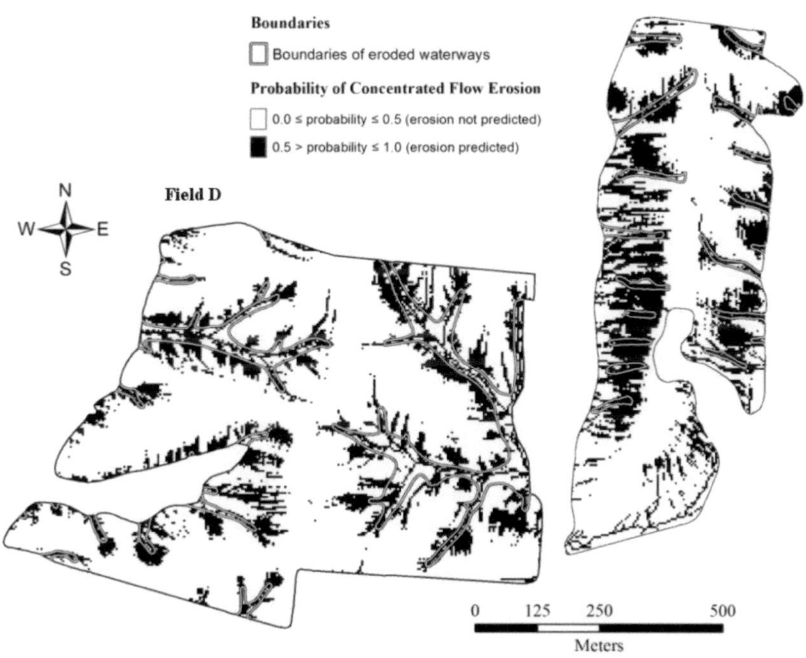

Figure 5. Discretized erosion probability maps derived from unsmoothed LiDAR data overlain by field boundaries. Terrain attributes were calculated with TauDEM using the D ∞ flow direction algorithm.

One problem with LiDAR data is that it contains systematic artifacts resulting from differential plant stubble heights in the direction perpendicular to crop rows associated with the direction of movement of farm machinery. With terrain modeling, they can create artificial flow lines along pathways of farm equipment. These lines are apparent in the unsmoothed but not smoothed LiDAR slope data (Figure 6). Unfortunately, the method of smoothing presented in this chapter introduced new artifacts along the contours because the procedure involved

1. calculating contours from DEMs and
2. rasterizing the contours.

This smoothing is very computationally resource intensive, and potentially problematic for professional conservation planners. It may be necessary for GIS experts to smooth the LiDAR data during preprocessing and make the analyses available to planners. More computationally efficient smoothing techniques should be considered that minimize new artifacts.

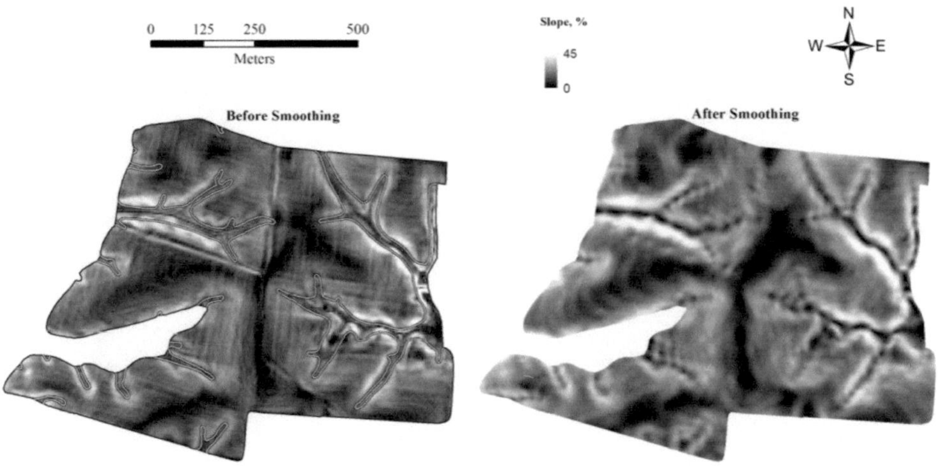

Figure 6. Comparison of unsmoothed and smoothed LiDAR slope data for Field D.

3.3. Comparison of LiDAR with RTK and USGS

The type-1 and -2 average misclassification errors for TauDEM D8 and D∞ output for LiDAR (8%) were similar in size as errors for the RTK (8%) dataset and lower than those for the USGS data (12%). The predictions were not very different for any of the datasets as shown in Figure 7. In many areas throughout the United States, LiDAR is being purchased for various applications (e.g., agriculture, soil mapping, transportation, land use planning). In those situations, it would be advantageous to use the LiDAR data for identifying eroded waterways. In areas where LiDAR is not available, USGS 10-m grids may be adequate for conservation planning [6].

3.4. Impact of Flow Direction Algorithm

The FD8 and DEMON methods were not used with the LiDAR data because TAPES no longer works with the latest version of ArcGIS (i.e., version 10.0). The regression parameters and Wald Chi Square test for the RTK (Table 4) and USGS (Table 5) analyses could be compared with those for the LiDAR models (Table 2). All of the parameters were statistically significant except for plan curvature in the USGS FD8 analyses (Table 5) which was consistent with [6]. The concentrated flow prediction models can be tested in other areas across the country with the same data source (RTK, LiDAR, and USGS) and flow direction (D8, D∞, FD8, and DEMON). The users should note that use of these may only be valid when similar smoothing techniques and DEM resolutions are used. In our case, all of our analyses were made with 4 by 4-m rasters. Grid scale is a very important factor to consider because at a small scale, there may be no relationship between land forms and curvature values; however, at a large increment, landscapes could have a profoundly large impact on an analysis similar to the one used in this study.

Flow Direction Method	Variable	Parameter Estimate	Wald Chi Square	
D8	Intercept	-3.14	57	**
	Topographic Wetness Index	0.251	18	**
	Length-Slope	0.681	42	**
	Plan Curvature	-7.979	188	**
D∞	Intercept	-6.17	119	**
	Topographic Wetness Index	0.603	65	**
	Length-Slope	1.01	77	**
	Plan Curvature	-5.73	82	**
FD8	Intercept	-9.88	161	**
	Topographic Wetness Index	0.882	88	**
	Length-Slope	1.88	137	**
	Plan Curvature	-5.54	39	**
DEMON	Intercept	-5.94	79	**
	Topographic Wetness Index	8.510	34	**
	Length-Slope	1.06	70	**
	Plan Curvature	-10.1	161	**

Table 4. Logistic regression parameters for the RTK dataset using the D8, D∞, FD8, and DEMON flow direction models.

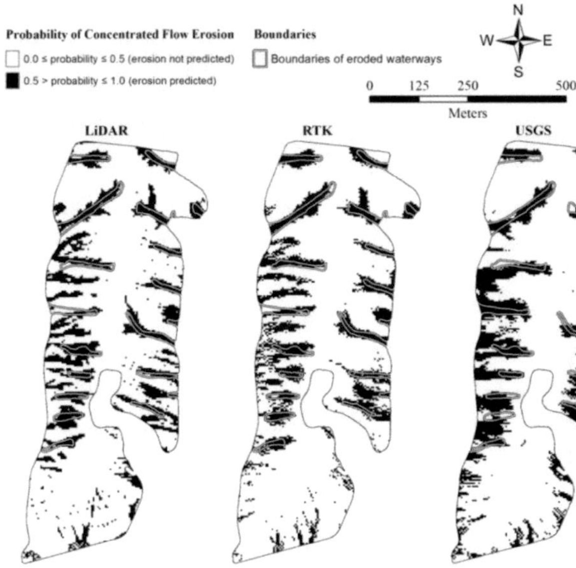

Figure 7. Discretized erosion probability maps for the logistic regression analysis of the LiDAR, RTK, and USGS datasets using the D∞ flow direction algorithm.

Flow Direction Method	Variable	Parameter Estimate	Wald Chi Square	
D8	Intercept	-3.08	87	**
	Topographic Wetness Index	0.351	56	**
	Length-Slope	0.323	17	**
	Plan Curvature	-4.46	99	**
D∞	Intercept	-7.19	157	**
	Topographic Wetness Index	0.829	123	**
	Length-Slope	0.790	70	**
	Plan Curvature	-2.11	19	**
FD8	Intercept	-7.63	200	**
	Topographic Wetness Index	0.827	147	**
	Length-Slope	1.02	124	**
	Plan Curvature	-0.565	2	ns
DEMON	Intercept	-8.23	123	**
	Topographic Wetness Index	0.919	96	**
	Length-Slope	0.895	72	**
	Plan Curvature	-5.69	60	**

Table 5. Logistic regression parameters for the USGS dataset using the D8, D∞, FD8, and DEMON flow direction models.

The average Type 1 and Type 2 error rates for the RTK dataset were ranked in the follow order (from lowest to highest): FD8 (6%), DEMON (7%), D8 (9%), and D∞(11%). The differences were smaller for the USGS dataset but procedures were ranked in a similar order: FD8 (10%), DEMON (10%), D∞(11%), D8 (12%). The map analysis for the RTK data (Figure 8) demonstrates that differences between methods were small with the FD8 model showing least noise in the southern part of the field.

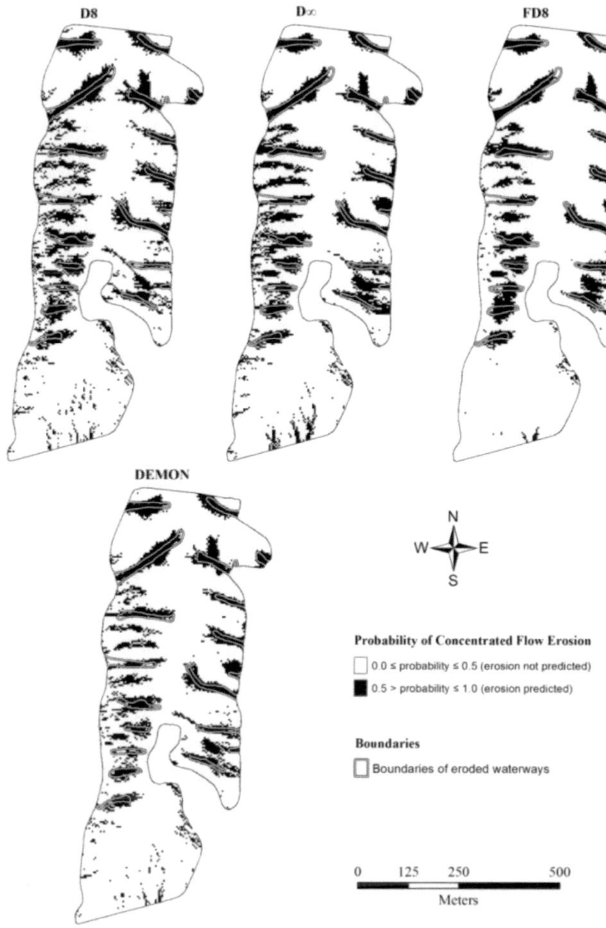

Figure 8. Discretized erosion probability maps for the logistic regression analysis of the RTK dataset comparing the D8, D∞, FD8, and DEMON flow direction algorithm.

4. Conclusion and Recommendations

The findings of this study indicate that LiDAR data can be used to clearly identify eroded features in agricultural landscapes with a level of accuracy that is similar to RTK GPS and better than USGS DEMs. It is critical that LiDAR data are smoothed prior to modeling erosion channels. Smoothing removes artifacts resulting from differences in plant residue heights perpendicular to the direction of travel by farm machinery. These differences can actually cause the terrain analysis flow models to incorrectly rout water along the direction of travel. While smoothing produced better results in this chapter, the method of smoothing with ArcGIS TopoToRaster was not efficient and cannot be used over large areas. More work is needed to determine computationally efficient smoothing algorithms for terrain analysis that minimize artifacts.

Conservation planners and GIS analysts should be able to accurately identify erosion features with the D8, D∞, FD8, and DEMON flow direction algorithm. This is important because previous work was based on TAPES G which is no longer being supported. Further TauDEM can utilize very large blocks of memory to cover extensive land areas, and can also operate on high performance computers. It is important to note that the choice flow algorithm will change the model parameters so it is important that they use the correct model. We recommend the TauDEM software program which uses the D8 and D∞ procedures because it works on 64 bit machines, allows the use of multiple core processer, and works with DEMs up to 4 GB in size.

All analyses performed in this study were based on with 4-m DEMs. Efforts are necessary to better understand the impact of the scale of terrain models on the quality of erosion model predictions. It may also be possible to expand the inference space of these models by including erosion parameters in the analyses obtained from soil surveys.

5. Nomenclature

D8, deterministic eight-neighbor; FD8, fractional deterministic eight-neighbor; DEMON, digital elevation model networks; DEM, Digital Elevation Model; TauDEM, Terrain Analysis Using Digital Elevation Models; TAPES, Grid-Based Terrain Analysis Programs for the Environmental Sciences; GIS, geographic information systems; RTK, Real Time Kinematic GIS, Geographic Information System.

Acknowledgements

We appreciate the generosity of Mike Ellis in providing us access to his grassed waterway datasets and Mike, Bob, and Jim Ellis for allowing us to conduct this research on their farm. We are also grateful for the assistance of Randall Rock, Jack Kuhn, Danny Hughes from the NRCS. We would like to thank Steve Workman from the UK college of Agriculture for support from the SB-271 Water Quality Funding. We also wish to express our appreciation to Photo Science president Mike Ritchie for providing us the complementary LiDAR data used in this chapter.

http://dx.doi.org/10.5772/51526

Author details

Adam Pike[1], Tom Mueller[2*], Eduardo Rienzi[2], Surendran Neelakantan[2], Blazan Mijatovic[2], Tasos Karathanasis[2] and Marcos Rodrigues[3]

*Address all correspondence to: mueller@uky.edu

1 Photo Science, Lexington, KY, USA

2 Department of Plant and Soil Sciences, University of Kentucky, Lexington, KY, USA

3 Univ. Estadual Paulista (UNESP), Jaboticabal, Brazil

References

[1] Thorne, C. R., Zezenbergen, L. W., Grissinger, E. H., & Murphey, J. B. (1986). Ephemeral Gullies as Sources of Sediment. *In: Proceeding of the 4th Federal Interagency Sedimentation Conference*, 24-27 March 1986, Las Vegas, United States. Washington, DC, United States: Gov. Print. Office.

[2] Moore, I. D., Burch, G. J., & Mackenzie, D. H. (1988). Topographic Effects on the Distribution of Surface Soil Water and the Location of Ephemeral Gullies. *Transactions of the ASABE*, 31(4), 1098-1107.

[3] Srivastava, K. P., & Moore, I. D. (1989). Application of Terrain Analysis to Land Resource Investigations of Small Catchments in the Caribbean. *In: Proceeding of the 20th Int. Conf. of the Erosion Control Association*, 15-18 Feb. 1989, Vancouver, BC, Canada. Streamboat Springs: International Erosion Control Association.

[4] Berry, J. K., Delgado, J. A., Pierce, F. J., & Khosla, R. (2005). Applying Spatial Analysis for Precision Conservation across the Landscape. *Journal of Soil and Water Conservation*, 60(6), 363-370.

[5] Pike, A. C., Mueller, T. G., Schörgendorfer, A., Shearer, S. A., & Karathanasis, A. D. (2009). Erosion Index Derived from Terrain Attributes using Logistic Regression and Neural Networks. *Agronomy Journal*, 101(5), 1068-1079.

[6] Pike, A. C., Mueller, T. G., Schörgendorfer, A., Luck, J. D., Shearer, S. A., & Karathanasis, A. D. (2010). Locating Eroded Waterways with United States Geologic Survey Elevation Data. *Agronomy Journal*, 102(4), 1269-1273.

[7] Luck, J. D., Mueller, T. G., Shearer, S. A., & Pike, A. C. (2010). Grassed Waterway Planning Model Evaluated for Agricultural Fields in the Western Kentucky Coal Field Physiographic Region of Kentucky. *Journal of Soil and Water Conservation*, 65(5), 280-288.

[8] Hodgson, M. E., Jensen, J. R., Schmidt, L., Schill, S., & Davis, B. (2003). An Evaluation of LiDAR and IFSAR-Derived Digital Elevation Models in Leaf-On Conditions with USGS Level 1 and Level 2 DEMs. *Remote Sensing of Environment*, 84(2), 295-308.

[9] Evans, M., & Linsay, J. (2010). High Resolution Quantification of Gully Erosion in Upland Peatlands at the Landscape Scale. *Earth Surface Processes and Landforms*, 35(8), 876-866.

[10] Perroy, R. L., Bookhagen, B., Asner, G. P., & Chadwick, O. A. (2010). Comparison of Gully Erosion Estimates Using Airborne and Ground-Based LiDAR on Santa Cruz Island, California. *Geomorphology*, 118(3-4), 288-300.

[11] O'Callaghan, J. F., & Mark, D. M. (1984). The Extraction of Drainage Networks from Digital Elevation Data. *Computer Vision, Graphics, and Image Processing*, 28(3), 323-344.

[12] Gallant, J. C., & Wilson, J. P. (2000). Primary Topographic Attributes. In: Wilson JP., Gallant JC. (ed.), *Terrain Analysis: Principles and Applications*, New York: John Wiley & Sons, 51-85.

[13] Moore, I. D., Gessler, P. E., Nielsen, G. A., & Peterson, G. A. (1993). Soil Attribute Prediction Using Terrain Analysis. *Soil Science Society of America Journal*, 57(2), 443-452.

[14] Lea, N. J. (1992). An Aspect-Driven Kinematic Routing Algorithm. In: Parsons AJ., Abrahams AD. (ed.), *Overland Flow: Hydraulics and Erosion Mechanics*, London: UCL Press, 393-407.

[15] Costa-Cabral, M. C., & Burges, S. J. (1994). Digital Elevation Model Networks (DEMON): a Model of Flow Over Hillslopes for Computation of Contributing and Dispersal Areas. *Water Resources Research*, 30(6), 1681-1692.

[16] Alexander, R. B., Smith, R. A., Schwarz, G. E., Boyer, E. W., Nolan, J. V., & Brakebill, J. W. (2008). Differences in Phosphorus and Nitrogen Delivery to the Gulf of Mexico from the Mississippi River Basin. *Environmental Science Technology*, 42(3), 822-830.

[17] EPA. (2008). Kentucky water quality assessment report. http://iaspub.epa.gov/waters10/attains_state.control?p_state=KY&p_cycle=2008, accessed 22 June 2012.

[18] Bilotta, G. S., & Brazier, R. E. (2008). Understanding the Influence of Suspended Solids on Water Quality and Aquatic Biota. *Water Research*, 42(12), 2849-2861.

[19] Bennett, S. J., Casali, J., Robinson, K. M., & Kadavy, K. C. (2000). Characteristics of Actively Eroding Ephemeral Gullies in an Experimental Channel. *Transactions of the American Society of Agricultural Engineers*, 43(3), 641-649.

[20] Fiener, P., & Auerswald, K. (2003). Effectiveness of Grassed Waterways in Reducing Runoff and Sediment Delivery from Agricultural Watersheds. *Journal of Environmental Quality*, 32(3), 927-936.

[21] Fiener, P., & Auerswald, K. (2009). Effects of Hydrodynamically Rough Grassed Waterways on Dissolved Reactive Phosphorus Loads Coming from Agricultural Watersheds. *Journal of Environmental Quality*, 38(2), 548-559.

[22] Soil Conservation Service. (1980). Soil survey of Shelby County, KY. USDA, NRCS, Fort Worth, TX.

[23] Mueller, T. G., Hartsockc, N. J., Stombaugh, T. S., Shearerb, S. A., Cornelius, P. L., & Barnhisel, R. I. (2003). Soil Electrical Conductivity Map Variability in Limestone Soils Overlain by Loess. *Agronomy Journal*, 95(3), 496-507.

[24] Pike, A. C., Mueller, T. G., Mijatovic, B., Koostra, B. K., Poulette, M. M., Prewitt, R. M., & Shearer, S. A. (2006). Topographic Indices: Impact of Data Source. *Soil Science*, 171(10), 800-809.

[25] Trimble. (1997). 4800 specifications. *Trimble Navigation Ltd., Sunnyvale, CA.*, Available online at, http://www.adwr.state.az.us/AzDWR/Hydrology/Geophysics/documents/Trimble4800Specs.pdf, accessed 22 June 2012.

[26] Trimble. (2008). Trimble 5800 GPS System Datasheet. *Trimble Navigation Ltd., Sunnyvale, CA.*, Available online at, http://trl.trimble.com/docushare/dsweb/Get/Document-32185/022543-016E_5800_GPS_DS_0708_LR.pdf, accessed 22 June 2012.

[27] Moore, I. D., & Wilson, J. P. (1992). Length-Slope Factors for the Revised Universal Soil Loss Equation: Simplified Method of Estimation. *Journal of Soil and Water Conservation*, 47(5), 423-428.

[28] Wilson, J. P., & Gallant, J. C. (2000). Digital Terrain Analysis. In: Wilson JP., Gallant JC. (ed.), *Terrain analysis: Principles and Applications*, New York: John Wiley & Sons, 1-28.

Chapter 4

Soil Erosion After Wildfires in Portugal: What Happens When Heavy Rainfall Events Occur?

L. Lourenço, A. N. Nunes, A. Bento-Gonçalves and A. Vieira

Additional information is available at the end of the chapter

http://dx.doi.org/10.5772/50447

1. Introduction

In Portugal, as well as in other Mediterranean countries, wildfires and burnt areas have increased significantly since 1970. This rising trend, although encompassing some periods of lower burnt areas, distinguishes Portugal from other southern European States with the highest number of ignitions and the greatest proportion of burnt areas, particularly in the central and northern regions (Nunes, 2012). Forest fires therefore constitute one of the most significant environmental problems (Moreno, 1989; Vallejo, 1997) and are frequently considered the major cause of soil degradation and desertification (Rubio, 1987).

Wildfires can considerably change hydrological processes and the landscape's vulnerability to major flooding and erosion events (Shakesby and Doerr, 2006; Stoof et al., 2012). Post-fire mudflows and flash floods represent a particularly acute problem in mountainous regions (Tryhorn et al., 2007). In fact, vegetation cover is an important factor in determining runoff and erosion risk (Nunes, 2011). Its removal by fire increases the raindrop impact on the bare soil and reduces the storage of rainfall in the canopy, thus increasing the amount of effective rainfall. Burned catchments are therefore at increased hydrological risk and respond faster to rainfall than unburned catchments (Meyer et al. 1995; Cannon et al. 1998; Wilson, 1999; Stoof et al., 2012). Wildfires also affect the hydrogeological response of catchments by altering certain physical and chemical characteristics of the soils, including their water repellent conditions (Conedera et al. 1998; DeBano et al. 1998; Letey 2001; Martin and Moody 2001; Shakesby and Doerr 2006). Increased runoff can lower the intensity threshold and the amount of precipitation needed to cause a flood event and also exacerbate the impact of precipitation. Combined with steep slopes, this can create the potential for flash floods.

Various studies in different parts of the world, including Portugal, have shown strong and sometimes extreme responses in runoff generation and soil loss following fires, especially during the earlier stages of the so-called "window-of-disturbance" (Shakesby, 2011).

In general, the first 4–6 months after a fire is often the period of greatest vulnerability to erosion because of the maximum fire potential in summer (July–August) and the likelihood of intense post-wildfire rainfall the following autumn–winter (November–January) (Sala et al., 1994; Andreu et al., 2001). However, soil erosion may reach its peak during the first year after a wildfire and subsequently decline, or in some situations be delayed until later, (much later in some cases) during the window of disturbance, in the third or even the fifth year after a fire (Mayor et al., 2007; Llovet et al., 2009). As noted by Ferreira et al. (2009), since the greatest effects of fire on hydrology and erosion generally occur shortly after a fire, data analysis and discussion is limited to the short-term (±1 yr) effects.

Post wildfire hydrological and erosional responses have been assessed at plot and hill slope level in various parts of the world, especially in the Mediterranean region, under natural rainfall conditions (Lourenço, 1989; Sala et al., 1994; Ferreira et al., 1997; Andreu et al., 2001, Coelho et al., 2004; Shakesby and Doerr, 2006).

The hydrogeomorphic responses to wildfire at catchment level have received much less attention than those on smaller scales in locations worldwide, mainly because of the greater practical difficulties and expense involved in monitoring on this scale, and the large chance factor involved in the wildfire burning even a small catchment completely (Shakesby and Doerr, 2006; Shakesby et al., 2006; Shakesby, 2011).

Despite the high rate of occurrences of fires in the European Mediterranean area (Moreira et al., 2001; Pausas, 2004), catchment-scale wildfire studies have mostly been carried out in the USA (Moody et, 2008; Moody and Martin, 2001; Gottfried et al., 2003; Meixner and Wohlgemuth, 2003; Nasseri, 1989; Seibert et al., 2010), South Africa (Scott and Van Wyk, 1990; Scott, 1993, 1997) and Australia (Brown, 1972; Langford, 1976; Prosser and Williams, 1998), and in only a few locations in the European Mediterranean area (Lavabre et al., 1993; Mayor et al., 2007; Ferreira et al, 2008; Stoof et al., 2012). In addition, post-fire monitoring is generally comparatively brief (usually 2–3 years) due to logistical and financial constraints, meaning that infrequent severe storms may be missed and the full recovery to pre-fire conditions may not be monitored.

Therefore, the impact of burned areas on peak flow and sediment transport in large river catchments has not been fully studied, although it is of the utmost importance to understand the off-site impacts of forest fires (Ferreira et al., 2008). A better understanding of the hydrogeomorphic impacts of fire at catchment level can improve our ability to understand, and therefore possibly predict, the risk of flooding and erosion in burned areas. In fact, when a precipitation event follows a large, high-severity fire, the impacts can cause various kinds of damage on- and off-site including high sediment inputs, downstream flooding, destruction of the aquatic habitat, and damage to human infrastructures.

Moreover, in the Mediterranean region precipitation patterns are highly variable in terms of time, space, amount and duration of events (Durão et al., 2010). The occurrence of heavy, often localised, precipitation can cause severe post-fire erosion and increase the risk of flash flooding and debris flow.

The main objective of this work was to evaluate the impact of fire at catchment level, with particular reference to the implications of the off-site hydrological response and erosional processes after severe rainstorms (involving one occurrence in June 2006 and another in July 2006). In fact, the growing probability of catastrophic wildfires in Portugal and elsewhere in the world has increased the need to understand the flood risk and the erosion and depositional responses of burned watersheds.

2. Study area

Two catchments (the Pomares and Piodão basins), both located in the mountains of central Portugal, were studied (Figure 1). The study area has a high annual precipitation rate, with an average of 1600/1700 mm yr^{-1}. The rainfall is generally concentrated during the period from October to May, whereas July and August are dry months. According to the Köppen climate classification, it has a Mediterranean Csb type climate.

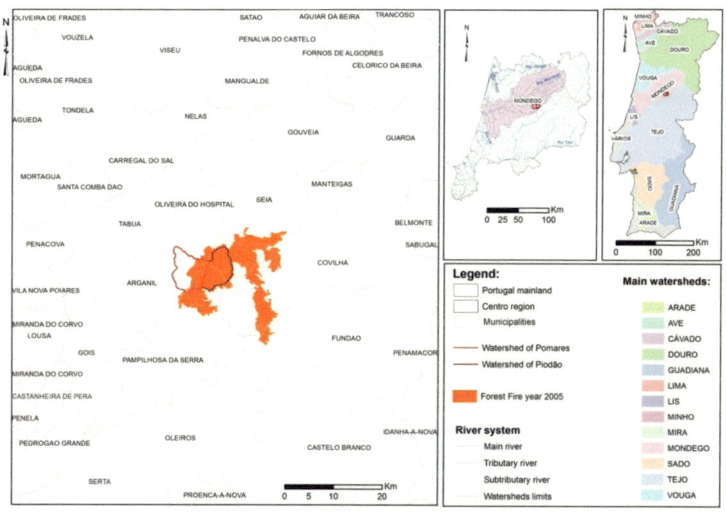

Figure 1. Location of the study basins and the areas affected by forest fire of 2005.

Both catchments lie on Precambrian schist and have shallow, stony, umbric leptosol soils. Both rivers are tributaries of River Alva and, according to the Strahler classification, are five-order streams. Some of the characteristics of both basins are presented in Table I. The Piodão and Pomares basins have areas of 34.3 and 44.7 km^2 respectively and both have a high elevation gradient of over 1,000 metres. In general, both are surrounded by steep slopes with a top convexity and no basal concavity. More than 90% of the basin areas have slopes of over

20% and in the Piodão river more than a half of the watershed has slopes of over 50%. A comparative analysis shows the basin ruggedness and coefficient of torrentiality to be slightly higher in the Pomares basin.

	Piodão river	Pomares river
Basin area (km2)	34.3	44.7
Basin gradient (m)	1047 (295-1342)	1069 (211-1280)
Basin ruggedness[1]	1.13	1.84
Drainage density[2] (km /km^2)	4.13	4.42
Coefficient of torrentiality	29.48	41.39
Basin area with slopes greater than 20 percent	43.9	54.7
Basin area with slopes greater than 50 percent	52.4	38.6
Burnt area, June 2005 (in%)	100	60

1. maximum change in elevation within a basin, divided by the square root of the basin area (Melton, 1965);2. the total length of all channels within a basin, divided by the basin area, (Horton, 1945)

Table 1. Main characteristics of both basins.

Important demographic and socio-economic changes have affected the mountain areas of Portugal for at least the last five to six decades. The population of the mountain areas decreased substantially during the second half of the 20th century, leading to the abandonment of agricultural land and a reduction in the size of herds and the amount of forest fuels consumed by grazing and the collection of firewood (Rego, 1992; Moreira et al., 2011; Lourenço, 1996, Nunes, 2012).

Consequently, the landscape has been drastically modified due to the sequential abandonment of traditional land use throughout the second half of the 20th century. The increase in uncultivated land has led to a secondary vegetation succession and modification of the vegetation structure, favouring horizontal and vertical fuel continuity and a consequent increase in flammable biomass. The unmanaged accumulation of large quantities of fuel and the exclusion of fires from forest management has led to a dramatic increase in the magnitude and frequency of forest fires (Carvalho et al., 2002; Moreira et al., 2011).

In addition, afforestation has focused primarily on highly inflammable species, mainly pines (predominantly *Pinus pinaster*) which also favours the proliferation of forest fires (Shakesby et al., 1996). Once fires break out under these highly dangerous conditions, they spread more easily and cannot be stopped. The low population density, delays in detecting fires, and difficulties in gaining access to the sites where fires tend to start, due to the rugged topography, are other factors that explain the large burnt areas in the central mountain area of Portugal.

The Mediterranean characteristics of the Portuguese climate (warm, dry summers and relatively wet winters) make it prone to wildfires and post-fire soil erosion. In Portugal, the major fires occur in summer, essentially in July and August. At this time of year, several factors combine to create the right conditions for the onset and propagation of wildfires. It is the driest time of year as well as the season for tourism, which includes camping and picnicking, and it is also the time when agricultural refuse and slash are traditionally cleaned and burned after crops have been harvested.

Consequently, as in other Mediterranean countries, Portugal's burnt area has increased significantly in recent decades. In the past three decades, the number of forest fires exceeded half a million ignitions and the total burnt area was approximately 3,236,890 ha, representing more than a third of the surface area of mainland Portugal (Nunes, 2012). Within the last 30 years (1981-2010), 2003 and 2005 were the worst fire seasons in Portugal, resulting in the burning of almost 430,000 hectares and 325.000 hectares respectively of forest land, shrub land and crops.

The Pomares and Piodão catchments have been severely affected by wildfires since the 1970s. Two large wildfires have affected the greater part of the area of both catchments: the first, between 13th and 20th September 1987, burnt a total of 10,900 hectares, and the second, occurring eighteen years later between 19th and 24th July 2005, affected an area of 17,450 hectares (Lourenço, 2006a b, 2007). Figure 1 shows the burnt area associated with both wildfires.

3. Methodology

The post-wildfire hydrological and erosional responses are based on intensive post-event fieldwork to determine the geomorphological impacts and socio-economic implications by collating, collecting and analysing data from field studies that was essential to understanding the meteorology, hydrology and hydraulics of the event.

The meteorological characteristics of the storms that affected the basins were determined using data from a rain gauge installed in the Piodão basin. Daily and 30-minute rainfall intensity measures (I_{30}) were chosen for each event, since rainfall frequency studies (Hershfield, 1961; Miller et al., 1973) indicate that in mountainous terrain 79% of the hourly rainfall occurs within 30 minutes and this type of storm has a short duration, lasting between 10 and 60 minutes (Moody and Martin, 2001).

The fieldwork took place a few days after the events occurred and was based on identifying certain variables:

Indicators of the peak discharge values, mainly cross-section surveys based on flood marks, in addition to signs of flow velocity (witness observations and water super-elevations in river bends or in front of obstacles). High water marks on channel banks, mostly indicated by the deposition of vegetation fragments and silt, were visible in the sites. These marks are very important and provide approximate estimates for reconstructing peak discharges for ungauged cross-sections of rivers affected by floods.

Sediment transfer processes (erosion and deposition on slopes and in river beds, hyperconcentrated mud or debris flow), which may give an indication of local runoff generation processes and flow energy and velocity.

The post-wildfire hydrological and erosional research benefited from the cooperation of local authorities and organisations that knew the area and had information about the catchment and the event. They provided useful information on the rainfall runoff processes (observation of surface runoff, origin of the runoff) and the local flow characteristics (type of flow – i.e. flood water, hyperconcentrated or debris flow, the presence of woody debris in the flow, approximate surface water flow velocities, blockages formed during the flood and their possible breakup, time and the effect of the collapse of bridges or dykes). The local authorities also provided important information on previous floods, which was relevant in assessing the return period of the flood.

After compiling the information using a Geographical Information System (GIS), detailed information was produced (mainly in the form of maps) which identified the areas heavily affected by water erosion (splash, rill and gully erosion) and sedimentation, as well as the areas affected by flash floods.

4. Results

4.1. The event of 16th June 2006

A rain gauge installed in the Piodão basin registered high levels of precipitation roughly one year after the July 2005 wildfire for two main events on 16th June and 14th July 2006. Figure 2 shows the 24 hour precipitation registered by the rain gauge during the month of June, totalling 58mm, distributed over 5 days (Figure 2).

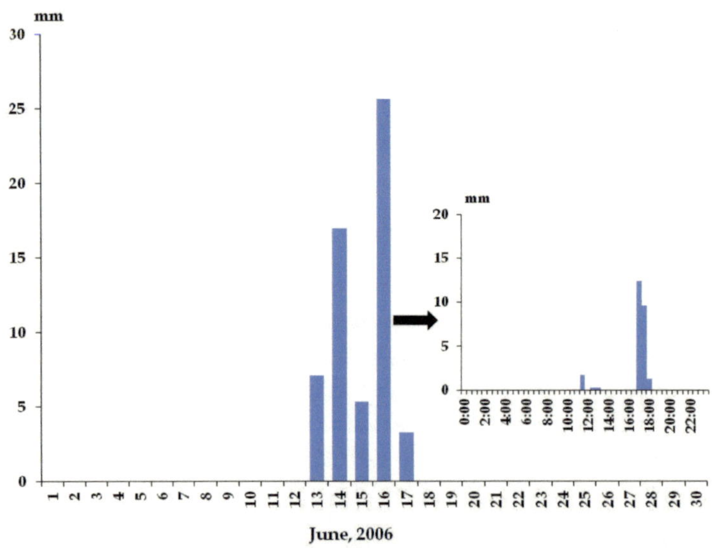

Figure 2. Daily distribution of rainfall in June and bi-hourly distribution on 16th June.

http://dx.doi.org/10.5772/50447

However, around 50% of the total rainfall was concentrated on 16th June. A more detailed analysis of the hourly distribution of rainfall on that day shows that 22 mm were recorded between 5 pm and 6 pm.

This event was caused by a high altitude cyclone in the southwest of the Iberian Peninsula which affected the weather in the Portuguese mainland during this period. In mid-latitudes, a 'cyclone' refers to the low pressure centres formed by baroclinic instability, with a typical scale in the order of 1000 km. However, cyclones or cyclonic centres also include any kind of surface depression, even small, weak, shallow low centres of orographic or thermal origin.

Following the high concentration of precipitation recorded on 16th June, several areas in both basins were affected by flash floods, soil erosion and sedimentation processes. Figure 3 summarises the areas worst affected by these processes.

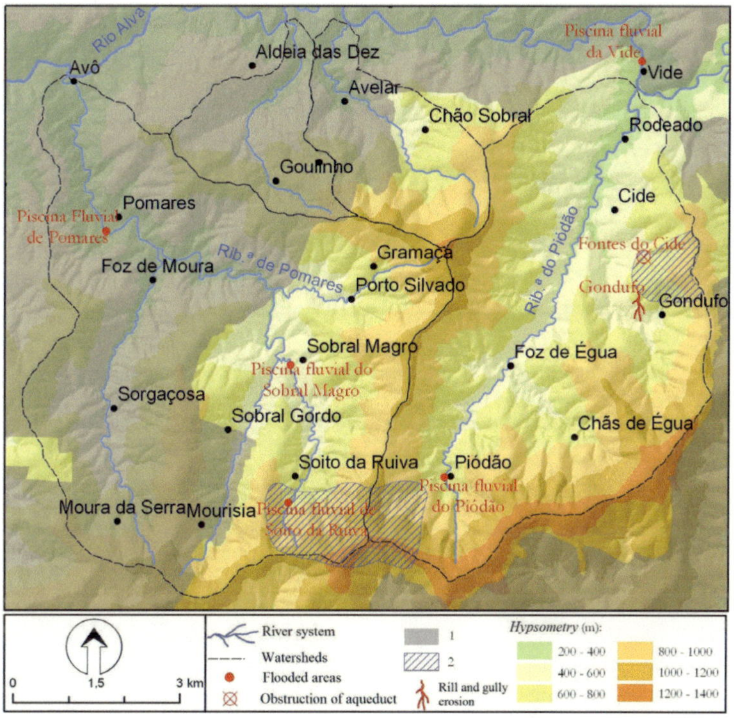

Figure 3. Effects of the intense rainfall after the wildfires.1. Area of the basin not affected by the wildfire of 2005; 2. Areas worst affected by the intense rainfall; Piscina fluvial="river beaches".

The figures 4 and 5 confirm the super-elevation of the flow at the Pomares Bridge (in the Pomares river basin) as well as the flooding of the right bank of the river. In fact, the stream flow created a 2.5 meter waterfront, although the floodgates were open. The impossibility of draining off the volume of water that had accumulated during the intense rainfall, as well as the power of the runoff and stream flow to transport materials obstructed the flow of the water and enlarged the flood area. Figure 5 simulates the peak discharge level and shows the tonnes of material, mainly branches of trees and shrubs, carried downstream, which created a blockage at the bridge.

Figure 4. The super-elevation of the flow at Pomares Bridge and the flooding of the right bank of the river.

Figure 5. Simulation of maximum peak discharge and the blocked organic and sediment debris (Pomares Bridge).

Upstream, at the of Sobral Magro and Soito da Ruiva river beaches the flood marks were also evident, as can be seen in figures and 6, 7 and 8. At Soito da Ruiva, the stream overflowed on both banks (Figure 6). In the Piodão basin the hydrological effects were also visi-

http://dx.doi.org/10.5772/50447

ble, particularly affecting the Piodão, Foz da Égua and Vide river beaches, where flash floods were recorded (Figure 3).

Figure 6. Simulation of the peak bank flood at Soita da Ruiva in the Pomares basin.

Figure 7. Deposition of sediment at Soito da Ruiva, in the Pomares basin.

During reconnaissance of the watersheds, widespread geomorphological consequences of the event were identified. In fact, high volume discharges have great erosional energy and the natural and man-made structures (dykes and bridges) along the rivers created obstacles to the transport of sediment and led to deposition throughout the main river channel and tributaries. However, the volume of off-site eroded sediment after a wildfire is difficult to assess because its response to rainstorms and runoff has different characteristics. The debris that was transported was mainly sediment from the thalwegs of tributaries that had been loosened by daily weathering and erosion, but could only be moved by large events.

Figures 7 and 8 show a plan of the debris flow deposition area caused by the inability of the drainage ditches to cope with the increased run-off generated in the upstream areas and the soil erosion, which led to flooding and the accumulation of large boulders and woody debris.

Figure 8. Wood accumulation following the wildfire at Sobral Magro, in the Pomares basin.

4.2. The event of 14th July 2006

In July, the precipitation was higher than the precipitation recorded in June, totalling 95 mm (Figure 9). This second event also registered very intensive rainfall. In fact, about 70mm fell in two days, on 13th and 14th July, registering 30mm and 39mm, respectively. The rainfall recorded on 14th July was concentrated in one single event that occurred between 4 pm and 5 pm. The total precipitation in the first half hour was 14mm, followed by 24mm in the next 30 minutes.

According to the Portuguese Meteorological Services, the heavy rainfall in several areas of inland Portugal was associated with "high atmospheric instability" related to the formation of a thermal low in the interior of the Iberian Peninsula, typical of the summer months. The summer heat in the Iberian Peninsula causes the surface pressure low (Alonso et al., 1994). If the Iberian thermal low draws air from the Atlantic rather than Africa, incursive winds can become humid, conditions become unstable, and intense thunderstorms may occur (Linés, 1977), sometimes leading to torrential rain.

According to Jarrett (2001), convective thunderstorms are known to have sharp rainfall gradients and rainfall intensities and vary in size, so that entire watersheds are not necessarily subjected to the same rainfall intensity.

The natural consequence of these precipitation patterns, which are relatively common in this climate, is that neighbouring watersheds receive different amounts of rainfall and therefore respond differently to the event. In fact, this event was more localised in comparison with the event of 16th June, mainly affecting the headwaters of the Piodão stream. The heavy rainfall significantly increased the amount of streamflow, resulting in a stronger and faster response and generating downstream floods and serious damage due to sediment transport. In addition to the substantial damage to human infrastructures, one death was recorded.

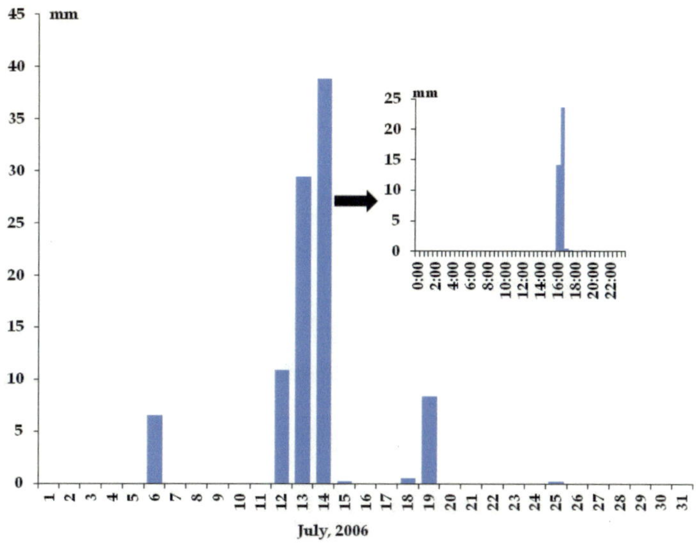

Figure 9. Daily distribution of rainfall in July and bi-hourly distribution on 14th July.

Figure 10, provided by a local resident and showing the volume of accumulated water, demonstrate that the peak discharge was higher during this event than the previous one. The flood marks on the house used to estimate the peak discharge level show that the ground floor was not flooded in the 16th June event, whereas during this flash flood the building was flooded to a depth of 1 metre.

The diagram in Figure 11 shows the longitudinal profile and different cross-sections of the Piodão river upstream of the village of Piodão, defines the stream bed and simulates the flood bed on the basis of flood marks, for the event of 14th July. Overall, the stream overflowed its banks and doubled in size in comparison to the "normal" bed. Immediately upstream of the village of Piodão, the flooded area was triple the size of the stream bed. This expansion of the flooded area was associated with a man-made structure designed as a channel for the bed stream. The inability to drain off the flow of water led to an increase in the flooded area, with profound geomorphologic consequences. The force of the water demolished a bridge which a tourist was crossing at the time, leading to his death. A car park was partially destroyed by the water, causing a landslide, as can be seen in figure 12.

Figure 10. Simulation of the maximum peak discharge in the 16th june (above) and in the 14th July (below). Comparative analysis.

In fact, intense rainfall increases the erosive power of overland flow, resulting in deeply incised channels, such as rills and gullies (figure 13), and accelerates the removal of material from hill slopes. Increased runoff can also erode significant volumes of material from channels. The net result of rainfall on burned basins is the transport and deposition of large volumes of sediment, both within and downstream of the burned areas. The following photographs illustrate its powerful capacity to transport materials along the main channel and its highly destructive force (Figures 14 and 15). In figure 14 a large block can be observed abandoned in the river bed. In figure 15 a trout pond is crammed with material transported by the flood. The power of the stream affected sediment transport processes during the flood, also influencing the morphology of the river.

5. Discussion

Wildfire is an important, and sometimes the most important, driving force behind landscape degradation in the Mediterranean region (e.g. Naveh, 1975; Andreu et al., 2001; Dimitrakopoulos and Seilopoulos, 2002; Alloza and Vallejo, 2006; Mayor et al., 2007). In fact, wildfire can have profound effects on a watershed. Burned catchments are at increased hydrological risk and respond faster to rainfall than unburned catchments (Meyer et al. 1995; Cannon et al. 1998; Ferreira et al. 2008; Stoof, 2012). Therefore, flooding and soil erosion also represent some of the most significant off-site impacts of wildfires, causing serious damage to public infrastructures and private property, as well as increased psychological stress for the affected population.

Wildfire alters the hydrological response of watersheds, including the peak discharge resulting from subsequent rainfall.

Peak discharge is also directly related to flood damage, and it is therefore important to understand the relationship between rainfall and peak discharge. The analysis of rainfall-runoff relations suggests that in the case of burned watersheds a rainfall intensity threshold exists, implying a critical change in the behaviour of the hydrological response. This threshold has been estimated at around 10 mm h^{-1} (Krammes & Rice, 1963; Doehring, 1968; Mackay and Cornish, 1982; Moody and Martin, 2001). One of the main reasons for the existence of a critical threshold intensity could be the hill slope infiltration rate. Infiltration rates have been shown to decrease by a factor of two to seven after wildfires (Cerdà, 1998; Martin & Moody, 2001), meaning that post-fire rainfall intensities that exceed this infiltration rate and cause runoff may be lower than the pre-fire intensities required to produce a comparable runoff. Below approximately 10 mm h^{-1} the rainfall intensity may be below the average watershed infiltration rate, meaning that most of the rainfall infiltrates, with some transient runoff (Ronan, 1986) and some subsurface flow, which may either cause quickflow (Hewlett and Hibbert, 1967) in the channel or a lagged response. Above 10 mm h^{-1} the rainfall intensity may exceed the average watershed infiltration rate, so that the runoff is dominated by sheet flow, which produces flash floods. As an example, Martin and Moody (2001), consider if the rainfall intensity is 20 mm h^{-1}, the unit-area peak discharge response would be 27 times greater than the response if the rainfall-runoff relation had not exceeded the 10 mm h^{-1} threshold. The same authors consider that if the rainfall intensity is 55 mm h^{-1} the response will be 700 times greater.

The consumption of the rainfall-intercepting canopy and soil-mantling litter and duff, intensive drying up of the soil, combustion of soil-binding organic matter, and enhancement or formation of water-repellent soils are factors that reduce rainfall infiltration into the soil and significantly increase overland flow and runoff in channels. The removal of obstructions to flow, such as live and downed timber and plant stems, due to wildfire can increase the erosive power of the overland flow, accelerating the removal of material from hill slopes. Increased runoff can also erode significant amounts of material from channels. The net result of rainfall on burned basins is often the transport and deposition of large volumes of sediment, both within and downstream of the burned areas (Cannon et al., 2008; Cannon 2005).

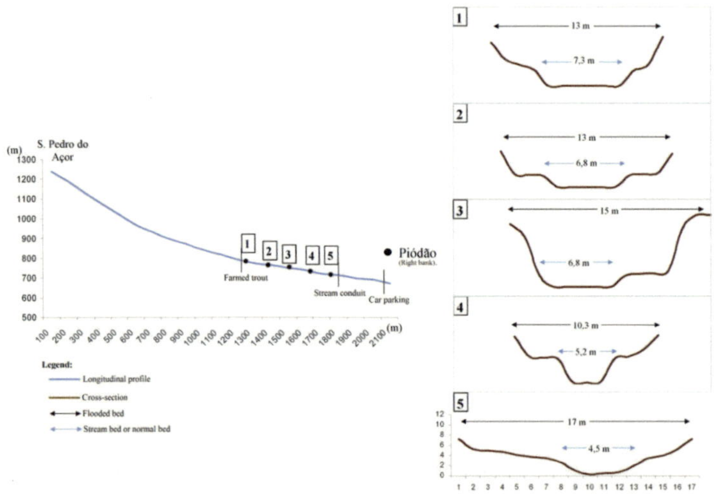

Figure 11. Profile and different cross-sections of the Piodão river upstream of the village of Piodão and normal and flooded area in the event of 14th July.

Figure 12. A car park partially destroyed by the water, causing a landslide.

Figure 13. Rills and gullies erosion as a consequence of the intense rainfall.

Post-fire debris flows are generally triggered by one of two processes: surface erosion caused by rainfall runoff, and landslides caused by the infiltration of rainfall into the ground. Runoff-dominated processes are by far the most common, since fires usually reduce the infiltration capacity of soils, which increases runoff and erosion. Infiltration processes are much less common, but prolonged heavy rain may increase soil moisture even after a wildfire. The wet soil may then collapse, producing infiltration-triggered landslides (Johnson, 2005).

According to (Johnson, 2005), although debris flows can occur in areas lying on almost any rock type, the areas most likely to produce debris flows are those lying on sedimentary or metamorphic rocks with more than around 65% of the area moderately or severely burned. In addition, debris flows are most frequently produced from steep (> 20), tightly confined drainage basins with an abundance of accumulated material, and are unlikely to extend beyond the mouths of basins larger than about 25 square kilometres (Johnson, 2005). The numerous instances of debris flows found in the study area suggest that the bedrock must have been highly fractured and weathered in order to be transported by the flow.

Figure 14. The powerful capacity to transport materials along the main channel during the event of 14th July.

Figure 15. A trout pond is crammed with material transported by the flood.

Despite the fact that the events studied occurred one year after the wildfires, it would be expected that the stream flow and erosion response would be much lower after vegetation regrowth and the removal of some of the sediment by relatively smaller storms in the following autumn and winter. Nevertheless, post-fire threshold conditions change over time even though the sediment supplies are depleted and the vegetation recovers, and the net result of intense rainfall on these burned basins was flash flooding in several areas and the transport and deposition of large volumes of sediment, both within and downstream of the burned areas. DeBano (2000) and Loaiciga (2001) consider that wildfires increase the magnitude of runoff and erosion and alter the hydrological response of watersheds resulting from subsequent rainfall, creating a risk for downstream communities that lasts for 1-3 years after a fire. Several other authors (Rowe et al., 1954; Doehring, 1968; Scott and Williams, 1978; Wells et al., 1979; Helvey, 1980; Robichaud et al., 2000) extend the "window of disturbance" to a much longer period of 3–10 years.

6. Conclusion

The hydrogeomorphic consequences of the 2006 events were identified during the field survey and it was found that there were widespread effects in the valleys of the watershed as well as in the main river channel and tributaries. In the Piodão and Pomares river basins, there were many instances of bed lowering, channel widening, avulsion and deposition. In several valleys there were flood marks, shallow landslides, slope failures and erosion gullies due to the intense rainstorm registered in both events. There were several instances of damage to infrastructures and buildings and one human life was lost.

Fires, floods and intensive erosion are a regular part of the landscape in mountainous regions around the world (Tryhorn et al., 2007) and are particularly significant in the Mediterranean basin, where forest fires have been increasing (JRC, 2005) and the climate is characterised by intense rainfall as a consequence of strong cyclogenesis (Kostopoulou, 2003). However, this intense rainfall has also been associated with factors other than cyclogenesis. Estrela et al. (2000) show that orographically induced thunderstorms caused by the Iberian thermal low can produce large volumes of precipitation. Post-fire floods may be associated with several different meteorological mechanisms and may either occur immediately after the fire or be delayed by several weeks or even years. Delayed floods are more likely to be caused by surface modifications that reduce infiltration, with precipitation due either to a large-scale drought break or localised thunderstorms. In combination, these processes can create a greater potential for severe flooding and intense erosive processes. A single intense rainstorm can generate peak flows which produce 75% of the sediment eroded during a longer (7-year) period of study (Shakesby, 2011).

In Portugal, several mountain areas have been affected by flash floods and landslides after forest fires. As an example, about 2 decades previously a major fire, which occurred in September 1987 and burnt an area of 10900 ha, affected most of the Pomares and Piodão basin area (Lourenço, 1988; 2006a b). A storm with similar characteristics to those in this study occurred in 2006, generating flash floods and severe erosion. Lourenço (1994) also studied a landslide which occurred in the Serra da Estrela mountains (in granitic lithology), after a severe rainfall event in October 1993, in an area burnt in August 1991. In the northern region of Portugal, Pedrosa et al. (2001) also studied a landslide that destroyed the great part of village of Frades (Arcos de Valdevez). This landslide also occurred in a granitic soil and was linked with a fire that occurred a few months before and destroyed the plant cover.

There is therefore a need to develop tools and methods to identify and quantify the potential hazards posed by flash floods and landslides generated by burned watersheds. An analysis of data collected from studies of flash flooding and debris flows following wildfires can answer many of the questions that are fundamental to post-fire hazard assessment—what and why, where, when, how big, and how often?

In fact, it is necessary to improve predictions of the magnitude and recurrence of the flooding that follows wildfires, due to the increased human population at risk in the wildland–urban interface. By understanding the magnitude of the runoff response and the erosion and deposition responses of recent wildfires, we can minimise loss of life and damage to property and provide data for landscape evolution in areas prone to wildfire. Moreover, watershed-scale predictions of erosion and deposition caused by these natural disasters can be used by land managers to prioritise forestry measures based on the erosion potential before and after wildfires.

Author details

L. Lourenço, A. N. Nunes[1], A. Bento-Gonçalves[2] and A. Vieira[2]

*Address all correspondence to: adelia.nunes@ci.uc.pt

1 Centro de Estudos em Geografia e Ordenamento do Território (CEGOT), Departamento de Geografia da Faculdade de Letras, Universidade de Coimbra Portugal

2 Centro de Estudos em Geografia e Ordenamento do Território (CEGOT), Departamento de Geografia, Universidade do Minho, Campus de Azurém Portugal

References

[1] Alloza, J. A., & Vallejo, V. R. (2006). Restoration of burned areas in forest management plans. *In: Desertification in the Mediterranean Region: a Security Issue, W.G. Kepner, J.L. Rubio, D.A. Mouat & F. Pedrazzini (eds)*, 475-488, 1402037597, Springer, Dordrecht.

[2] Alonso, S., Portela, A., & Ramis, C. (1994). First considerations on the structure and development of the Iberian thermal low-pressure system. *Ann. Geophys.*, 12, 457-468, 0992-7689.

[3] Andreu, V., Imeson, A. C., & Rubio, J. L. (2001). Temporal changes in soil aggregates and water erosion after a wildfire in a Mediterranean pine forest. *Catena*, 44, 69-84, 0341-8162.

[4] Brown, J. A. H. (1972). Hydrologic effects of a bushfire in a catchment in south-eastern New South Wales. *Journal of Hydrology*, 15, 77-96, 0022-1694.

[5] Cannon, S., Powers, P., & Savage, W. (1998). Fire-related hyperconcentrated and debris flows on Storm King Mountain, Glenwood Springs, Colorado, USA. *Environ. Geol.*, 35, 210-218, 0943-0105.

[6] Cannon, S. H., & Gartner, J. E. (2005). Wildfire-related debris flow from a hazards perspective. *In: Debris-flow hazards and related phenomena, O. Hungr & M. Jacob (eds.)*, 321-344, 978-3540207269, Springer, Praxis Books in Geophysical Sciences.

[7] Cannon, S. H. (2000, 16-18 August). Debris-flow response of southern California watersheds recently burned by wildfire. Paper presented at Proceedings of the Second International Conference on Debris-Flow Hazards Mitigation, Taipei, Taiwan. *In: Debris-Flow Hazards Mitigation- Mechanics, prediction, and assessment, G.F. Wieczorek & N.D. Naeser (eds.)*, 45-52, 905809149X, Rotterdam.

[8] Cannon, S. H. (2001). Debris-flow generation from recently burned watersheds. *Environmental and Engineering Geoscience*, 7, 321-341, 1078-7275.

[9] Cannon, S. H. (2005). Southern California-Wildfires and debris flows. *U.S. Geological Survey Fact Sheet*, 2005-3106, http//pubs.usgs.gov/fs/2005/3106.

[10] Cannon, S. H., Gartner, J. E., Wilson, R. C., & Laber, J. L. (2008). Storm rainfall conditions for floods and debris flows from recently burned areas in southwestern Colorado and southern California. *Geomorphology*, 96, 250-269, 0169-555X.

[11] Cannon, S. H., Kirkham, R. M., & Parise, M. (2001). Wildfire-related debris-flow initiation processes, Storm King Mountain, Colorado. *Geomorphology*, 39, 171-188, 0169-555X.

[12] Carvalho, T. M., Coelho, C. O. A., Ferreira, A. J. D., & Charlton, C. A. (2002). Land degradation processes in Portugal: Farmers' perceptions of the application of European agroforestry programmes. *Land Degradation & Development*, 13, 177-188, 1099-145X.

[13] Cerdà, A. (1998). Post-fire dynamics of erosional processes under Mediterranean climatic conditions. *Zeitschrift fuer Geomorphologie Neue Folge*, 42(3), 373-398, 0375-8109.

[14] Coelho, C. O. A., Ferreira, A. J. D., Boulet, A. K., & Keizer, J. J. (2004). Overland flow generation processes, erosion yields and solute loss following different intensity fires. *Quarterly Journal of Engineering Geology and Hydrogeology*, 37, 233-240, 1470-9236.

[15] Conedera, M., Marxer, P., Ambrosetti, P., Bruna, G. D., & Spinedi, F. (1998). The 1997 forest fire season in Switzerland. *Int. For. Fire News*, 18, 85-88, 1029-0864.

[16] DeBano, L. F., Neary, D. G., & Ffolliott, P. F. (1998). *Fire's Effects on Ecosystems*, 0-47116-356-2, Wiley & Sons, New York.

[17] DeBano, L. F. (2009, March/April). Fire Effects on Watersheds: an overview. *Southwest Hydrology*, 2, 1552-8383.

[18] Dimitrakopoulos, A. P., & Seilopoulos, D. (2002). Effects of rainfall and burning intensity on early post-fire soil erosion in a Mediterranean forest of Greece. *In: Proceedings of the Third International Congress. Man and Soil at the Third Millennium, J.L. Rubio, R.P.C Morgan, S. Asins & V. Andreu (Eds.)*, II, 1351-1357, 84-87779-45-X, Geoforma Ediciones, Logroño.

[19] Doehring, D. O. (1968). The effect of fire on geomorphic processes in the San Gabriel Mountains, California. *In: Contributions to Geology, R.B. Parker (Ed.)*, 43-65, Laramie, University of Wyoming, Wyoming.

[20] Durão, R. M., Pereira, M. J., Costa, A. C., Delgado, J., Del Barrio, G., & Soares, A. (2010). Spatial-temporal dynamics of precipitation extremes in southern Portugal: a geostatistical assessment study. *International Journal of Climatology*, 30(10), 1526-1537, 1097-0088.

[21] Linés, A. (1970). The climates of the Iberian Peninsula. *In: Climates of Northern and Western Europe. C. C. Wallén (Ed.)*, 0042-9767, World Meteorological Organization.

[22] Ferreira, A. J. D., Coelho, C. O. A., Ritsema, C. J., Boulet, A. K., & Keizer, J. J. (2008). Soil and water degradation processes in burned areas: lessons learned from a nested approach, *Catena*, 74, 273-285, 0341-8162.

[23] Ferreira, A. J. D., Coelho, C. O. A., Shakesby, R. A., & Walsh, R. P. D. (1997). Sediment and solute yield in forest ecosystems affected by fire and rip-ploughing techniques, central Portugal: a plot and catchment analysis approach. *Physics and Chemistry of the Earth*, 22, 309-314, 1474-7065.

[24] Helvey, J. D. (1980). Effects of a north central Washington wildfire on runoff and sediment production. *Water Resources Bulletin*, 16(4), 627-634, 0043-1370.

[25] Hewlett, J. D., & Hibbert, A. R. (1967). Factors affecting the response of small watersheds to precipitation in humid areas. *In: Forest Hydrology, Proceedings of a National Science Foundation Advance Science Seminar, WE Sopper & HW Lull (eds)*, 275-290, Pergamon Press, New York.

[26] Hershfield, D. M. (1961). Rainfall frequency atlas of the United States for duration from 30 minutes to 24 hours and return periods from 1 to 100 years. *US Department of Commerce, Technical Paper* [40], 0–938909–67.

[27] Jarrett, R. D. (1987). Errors in slope-area computations of peak discharges in mountain streams. *Journal of Hydrology*, 96, 53-67, 0022-1694.

[28] JRC. (2005). Forest fires in Europe 2004. Joint Research Center S.P.1.05.147 EN 978-9-27916-494-1 European Communities

[29] Johnson, M. L. (2005). Southern California-Wildfires and Debris Flows. *U.S. Department of the Interior, U.S. Geological Survey*, 4, 0196-1497.

[30] Kostopoulou, E. (2003). The relationships between atmospheric circulation patterns and surface climatic elements in the eastern Mediterranean. University of East Anglia, *Ph.D. thesis*, 407.

[31] Krammes, J. S., & Rice, R. M. (1963). Effect of fire on the San Dimas experimental forest In: Arizona Watershed Symposium, Proceedings 7th Annual Meeting, 0 471 74283 X Phoenix, Arizona , 31-34.

[32] Langford, K. J. (1976). Change in yield of water following a bushfire in a forest of eucalyptus regnans. *Journal of Hydrology*, 29(1-2), 87-114, 0022-1694.

[33] Lavabre, J., Torres, D. S., & Cernesson, F. (1993). Changes in the hydrological response of a small Mediteranean basin a year after a wildfire. *Journal of Hydrology*, 142, 273-299, 0022-1694.

[34] Letey, J. (2001). Causes and consequences of fire-induced soil water repellency. *Hydrological Processes*, 15, 2867-2875, 1099-1085.

[35] Loaiciga, H. A., Pedreros, D., & Roberts, D. (2001). Wildfire-streamflow interaction in a chaparral watershed. *Advances in Environmental Research*, 5, 295-305, 1093-7927.

[36] Lourenço, L. (1988). Tipos de tempo correspondentes aos grandes incêndios florestais ocorridos em 1986 no Centro de Portugal. *Finisterra*, 23, Lisboa, 251-270, 0430-5027.

[37] Lourenço, L., Pedrosa, A., & Felgueiras, J. (2001). Movimentos em massa. Exemplos ocorridos no Norte de Portugal, ENB. *Revista Técnica e Formativa da Escola Nacional de Bombeiros*, 17, Sintra, 25-39, 1645-0086.

[38] Lourenço, L. (1994, 15 a 17 de Dezembro). A Enxurrada do Ribeiro de Albagueira. *III Congresso Florestal Nacional, Figueira da Foz*, 1-9.

[39] Lourenço, L. (2006a). (Cood) Paisagens de Socalcos e Riscos Naturais em vales do Rio Alva, Colectâneas Cindínicas VI, Projecto Interreg III B/Sudoe-Terrisc, 9-72833-020-0 de Investigação Científica de Incêndios Florestais da Faculdade de Letras da Universidade de Coimbra, Lousã, 192 p.

[40] Lourenço, L. (2006b). (Cood). Bacias hidrográficas das ribeiras do Piódão e de Pomares (Concelho de Arganil). Terrisc- Recuperação de paisagens de socalcos e prevenção de riscos naturais nas serras do Açor e da Estrela, Relatório Técnico 0605, 9-72994-626-4 Interreg III B/Sudoe-Terrisc, Núcleo de Investigação Científica de Incêndios Florestais da Faculdade de Letras da Universidade de Coimbra, Lousã, Portugal

[41] Lourenço, L. (2007). (Cood). Riscos Ambientais e Formação de Professores (Actas das VI Jornadas Nacionais do Prosepe), Colectâneas Cindínicas VII, 9-72833-021-7 de Sensibilização e Educação Florestal e Núcleo de Investigação Científica de Incêndios Florestais e Faculdade de Letras da Universidade de Coimbra, Coimbra, Portugal

[42] Lourenço, L. (1989). Quantificação da erosão produzida na serra da Lousã na sequência de incêndios florestais. Resultados preliminares. Relatório Técnico nº9816, 972-9038-66X GMF, Coimbra.

[43] Lourenço, L. (1996). *Serras de Xisto do Centro de Portugal- Contribuição para o seu conhecimento geomorfológico e geo-ecológico. PhD*, Universidade de Coimbra, Portugal.

[44] Llovet, J., Ruiz-Valera, M., Josa, R., & Vallejo, V. R. (2009). Soil responses to fire in Mediterranean forest landscapes in relation to the previous stage of land abandonment. *International Journal of Wildland Fire*, 18, 222-232, 1049-8001.

[45] Mackay, S. M., & Cornish, P. M. (1982). Effects of wildfire and logging on the hydrology of small catchments near Eden, N.S.W. *In: The First National Symposium of Forest Hydrology*, 111-117, 82(6), 0-85825-175-2, Eng. Aust. Natl. Conf. Publ.

[46] Mayor, A. G., Bautista, S., Llovet, J., & Bellot, J. (2007). Post-fire hydrological and erosional responses of a Mediterranean landscape: seven years of catchment-scale dynamics. *Catena*, 71, 68-75, 0341-8162.

[47] Meixner, T., & Wohlgemuth, P. M. (2003, October 28-30). Climate variability, fire, vegetation recovery, and watershed hydrology. Benson, Arizona. *In: Proceedings of the First Interagency Conference on Research in the Watersheds*, 651-656, 978-1-40206-588-0.

[48] Meyer, G. A., Wells, S. G., & Jull, A. J. T. (1995). Fire and alluvial chronology in Yellowstone National Park: Climatic and intrinsic controls on Holocene geomorphic processes. *Geol. Soc. Amer. Bull.*, 107, 1211-1230, 1943-2674.

[49] Miller, J. F., Frederick, R. H., & Tracey, R. J. (1973). Precipitation frequency Atlas of the Western United States, Colorado. *NOAA Atlas 2*, III, Nat. Weather Ser., Washington, DC, 0093-7177.

[50] Moody, J. A., & Martin, D. A. (2001). Initial hydrologic and geomorphic response following a wildfire in the Colorado Front Range. *Earth Surface Processes and Landforms*, 26, 1049-1070, 1096-9837.

[51] Moody, J., Martin, D., & Cannon, S. (2008). Post-wildfire erosion response in two geologic terrains in the western USA. *Geomorphology*, 95, 103-118, 0169-555X.

[52] Moreira, F., Viedma, O., Arianoutsou, M., Curt, T., Koutsias, N., Rigolot, E., et al. (2011). Landscape e wildfire interactions in southern Europe: implications for landscape management. *Journal of Environmental Management*, 92, 2389-2402, 0301-4797.

[53] Nasseri, I. (1989). Frequency of floods from a burned chaparral watershed. In: Berg, Neil H., tech. coord. Proceedings of the symposium on fire and watershed management; 1988 October 26-28; Sacramento, CA. Gen. Tech. Rep. PSW-109. Berkeley, CA: U.S. Department of Agriculture, Forest Service, Pacific Southwest Forest and Range Experiment Station: 68-71. Available from: National Technical Information Service, Springfield, VA 22161; PB89-228639.

[54] Naveh, Z. (1975). The evolutionary sequence of fire in the Mediterranean region. *Vegetatio*, 29, 199-208, 0042-3106.

[55] Nunes, A. N. (2012). Regional variability and driving forces behind forest fires in Portugal an overview of the last three decades (1980-2009). *Applied Geography*, 34, 576-586, 0143-6228.

[56] Nunes, A. N. (2011). Soil erosion under different land use and cover types in a marginal area of Portugal. *In Godone, D. & Stanchi, S. (eds.) Soil Erosion Studies*, In Tech Open Access Publisher, 59-86, 978-9-53761-934-3, available in, http://www.intechweb.org/.

[57] Prosser, I. P., & Williams, L. (1998). The effect of wildfire on runoff and erosion in native Eucalyptus forest. *Hydrological Processes*, 12, 251-265, 1099-1085.

[58] Rego, F. C. (1992). Land use changes and wildfires. *In Responses of forest ecosystems to environmental changes A. Teller, P. Mathy, & J. N. R. Jeffers (Eds.)*, 367-373, 978-1851668786, Elsevier Applied Science, London.

[59] Robichaud, P. R., Mc Cool, D. K., Pannkuk, C. D., Brown, R. E., & Mutch, P. W. (2001). Trap efficiency of silt fences used in hillslope erosion studies In: Proceedings of the International Symposium, Soil Erosion Research for the 21st Century, AscoughII II, J.C., Flanaga, D.C. (Eds.), American Society of Agricultural Engineers , 7-03-003239-X Q.420, St. Joseph, Michigan , 541-543.

[60] Robichaud, P. R., Beyers, J. L., & Neary, D. G. (2000, September). Evaluating the effectiveness of postfire rehabilitation treatments. United States Department of Agriculture, Forest Service Rocky Mountain Research Station, *General Technical Report RMRS-GTR-63*, 89.

[61] Ronan, N. M. (1986). The hydrological effects of fuel reduction burning and wildfire at wallaby Creek. *Melbourne and Metropolitan Board of Works, Report No. MMBW-W-0015*, 0-72416-738-2.

[62] Rowe, P. B., Countryman, C. M., & Storey, H. C. (1954). *Hydrologic analysis used to determine effects of fire on peak discharge and erosion rates in southern California*, USDA Calif. Forest and range experiment station and University of California, Berkeley, CA, 49.

[63] Sala, M., Soler, M., & Pradas, M. (1994). Temporal and spatial variations in runoff and erosion in burnt soils. Coimbra, Portugal. *In: Proceedings of the Second International Conference on Forest Fire Research*, 1123-1134, 972-97406-0-7, II.

[64] Scott, D. F., & Van Wyk, D. B. (1990). The effects of wildfire on soil wettability and hydrological behaviour of an afforested catchment. *Journal of Hydrology*, 121, 239-256, 0022-1694.

[65] Scott, D. F. (1993). The hydrological effects of fire in South African mountain catchments. *Journal of Hydrology*, 150, 409-432, 0022-1694.

[66] Scott, D. F. (1997). The contrasting effects of wildfire and clearfelling o the hydrology of a small catchment. *Hydrological Processes*, 11, 543-555, 1099-1085.

[67] Seibert, J., Mc Donnell, J. J., & Woodsmith, R. D. (2010). Effects of wildfire on catchment runoff response: a modelling approach to detect changes in snow-dominated forested catchments. *Hydrology Research*, 41, 378-390, 0029-1277.

[68] Shakesby, R., & Doerr, S. (2006). Wildfire as a hydrological and geomorphological agent. *Earth-Science Reviews*, 74, 269-307, 0012-8252.

[69] Shakesby, R. A., Boakes, D., Coelho, C. O. A., Gonçalves, A. J. B., & Walsh, R. P. D. (1996). Limiting the soil degradation impacts of wildfire in Pine and Eucalyptus forest in Portugal. *Applied Geography*, 16, 337-355, 0143-6228.

[70] Shakesby, R. A. (2011). Post-wildfire soil erosion in the Mediterranean: Review and future research directions. *Earth-Science Reviews*, 105, 71-100, 0012-8252.

[71] Shakesby, R. A., & Doerr, S. H. (2006). Wildfire as a hydrological and geomorphological agent. *Earth Science Reviews*, 74, 269-307, 0012-8252.

[72] Stoof, C. R., Vervoort, R. W., Iwema, J., Elsen, E., Ferreira, A. J. D., & Ritsema, C. J. (2012). Hydrological response of a small catchment burned by experimental fire. *Hydrol. Earth Syst. Sci.*, 16, 267-285, 1027-5606.

[73] Tryhorn, L., Lynch, A., & Abramsom, A. (2007). On the Meteorological Mechanisms Driving Postfire Flash Floods: A Case Study. *Monthly Weather Review*, 136, 1778-1791, 0027-0644.

[74] Wells, W. G., II. (1981). Some effects of brushfires on erosion processes in coastal Southern California. *In: Erosion and Sediment Transport in Pacific Rim Steeplands* [132], 305-342, 0-94757-111-6, IAHS.

[75] Wilson, C. J. (1999). Effects of logging and fire on runoff and erosion on highly erodible granitic soils in Tasmania. *Water Resources Research*, 35(11), 3531-3546, 0043-1397.

Chapter 5

Modeling of Soil Erosion and Its Implication to Forest Management

Nuray Misir and Mehmet Misir

Additional information is available at the end of the chapter

http://dx.doi.org/10.5772/ 53741

1. Introduction

Conservation of natural forest ecosystems will require a land ethic as prelude to understanding the functioning of forest ecosystems, ecological and physiological impacts of disturbances on ecosystems, and the processes involved in recovery of disturbed ecosystems. Many of the harmful effects of pollution, fire, flooding, and soil compaction can be abated by judicious planning measurements to create and perpetuate the critical components of forest stand structure and species composition. Strategies for continuous production of the products and services that can be supplied by forest ecosystems will need to be reinforced by expanded long-term research and close cooperation among various disciplines such as forest biologists, social scientists, economists, and regulatory government agencies [17].

Nowadays, multi-objective planning is necessary in forestry because of increased and varied demand for forest products and services. Management objective such as production of quality potable water, carbon stocking, aesthetic, recreation and community health in forest especially adjacent to big cities are of great importance. Forests have managed to produce wood products at various diameters and quality classes as the society demanded overtime [24]. Afterwards, the importance of these objectives has gradually diminished and overwhelmed by other management objectives such as conservation of water resources, prevention of soil erosion, carbon stocking, creation of landscape aesthetic, camouflaging military facilities and allocation of land for recreation [2]. The forest values on be grouped as static and dynamic forest values (Figure 1).

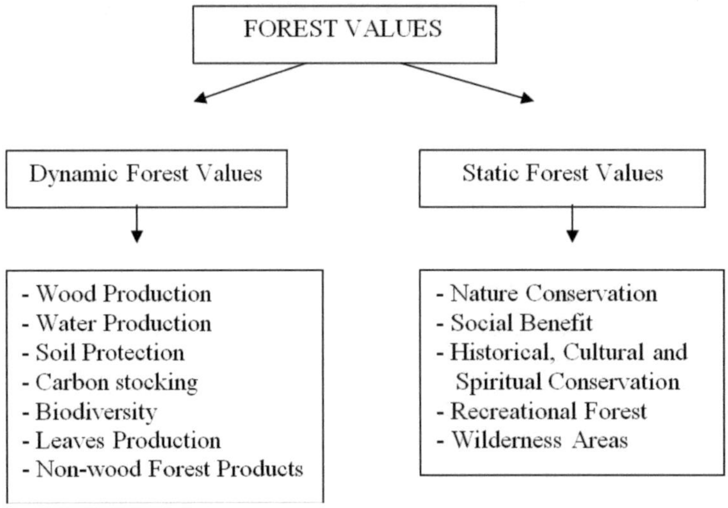

Figure 1. Classification of Forest values [23]

2. Soil Erosion

Erosion, the detachment of soil particles, occurs by the action of water, wind, or glacial ice. Such 'background' soil erosion has been occurring for some 450 million years, since the first land plants formed the first soil. Only erosion caused by water will be considered here. Water related erosion occurs when raindrops, spring runoff, or floodwaters wear away and transport soil particles. Erosion is a complex natural process that has often been accelerated by human activities such as land clearance, agriculture, construction, surface mining, and urbanization.

Soil erosion by water and wind affects both agriculture and the natural environment, and is one of the most important of today's environmental problems. It isn't easy to find comprehensive information about erosion, as the subobject is multidisciplinary involving geomorphologists, agricultural engineers, soil scientists, hydrologists and others; and is of interest to policy-makers, farmers, environmentalists and many other groups.

Schumm and Harvey [32] believe some of the terms used to describe erosion are misleading. Normal erosion and geologic erosion are often meant to imply pre-agricultural conditions of low erosion rates, whereas accelerated erosion and historic erosion imply greatly increased erosion rates caused by man. Because of the great variability in natural erosion both spatially and temporally at present and throughout geologic time, neither concept is correct. They prefer the term natural erosion for normal and geologic erosion, and the term man-induced erosion for accelerated and historic erosion.

3. Forest Management and Soil Erosion

All forest management activities affect soils, with effects ranging over a continuum from nearly none where the activity is minimal to large. To foster communication, a threshold should be established above which effects merit attention and below which further consideration is not justified. The magnitude of that threshold varies with the state of knowledge, about forest dynamics and must include recognition of uncertainty. Failure to identify thresholds inhibits communication to a wider audience and even among ourselves [12].

There are two kinds of effects of forest management on soils. The first, direct effect is an alteration of soil properties such as an increase in bulk density following passage of heavy equipment. Soil scientists generally agree on those direct effects; recognition of those alterations is literally axiomatic. The second effect of management on soils is indirect; a change in site productivity due to alteration of soil properties. Some of those secondary effects are obvious enough that can be considered corollaries. Specific studies and personal and vicarious experience have led to this worldview. Conversely, some of the indirect effects of management on soils are not as clear, and can be considered postulates. The distinction between axioms, corollaries, and postulates is often in the eye of the beholder, and depends on interpretation of both published reports and personal observations. Papers that support a position are evaluated differently than those in opposition. I offer no excuses for bias; "For every expert, there is an equal and opposite expert" [6].

Erosion is a natural process, but one whose rate and extent is exacerbated by forest management [36]. Most emphasis on erosion has been directed towards its effects on water quality and fish habitat, but because it involves displacement of soil, the growing medium, erosion also can affect site productivity [21]. However, forest management activities are necessary parts of forestry, and there may be minimal control over the circumstances under which they are carried out. Alterations of soil physical properties are extensive, immediate, and their effects in reducing productivity are well-documented. Chemical and biological properties of soils are also changed by management activities, but the effects on productivity are less well-documented and of longer term; their influence is not clear. Historical evidence shows that forest ecosystems are dynamic and resilient. Assessment of the consequences of changes in properties must recognize that shifts in preferred species should not be equated with changes in productivity, and that short-term effects, measured by the length of most experiments or observations, may not be indicative of long-term effects [12]. Accurate assessment of the effects of its change, however, is likely to continue to be obscured by the influence of the many other elements that also affect forest productivity [40]. At our current state of ignorance, a reasonable approach may be a simple sensitivity analysis that uses spatially based techniques (geographic information systems) and reasonable estimates of effects of the many factors that affect forest productivity to develop an impression of changes in soil productivity [12].

Use of more sophisticated simulation models implies greater knowledge than we currently possess. Both ethical and economic considerations demand good stewardship with professional accountability for our natural resources. Extensive forest management, if carried out with both wisdom and prudence, is not antithetical to good stewardship. "All of us have vested interests in making forest management a wise and efficient use of resources. Soil information can immeasurably help us be good stewards of the land" [12].

4. Soil Loss Characterization

Characterization of soil loss is very important for environment and natural resources. In erosion control planning, soil loss estimates for a particular site are determined using a prediction model and compared with a T-value for that site [31]. The Universal Soil Loss Equation (USLE) is an example of a model used extensively to predict erosion from croplands and rangelands. More recently, the Agricultural Research Service, Forest Service, and the Bureau of Land Management have joined in a cooperative effort, the Water Erosion Prediction Project (WEPP). WEPP has been implemented to develop an improved model based on modern technology for estimating soil erosion by water. WEPP technology is based on fundamental hydrologic and soil erosion processes and is designed to replace the widely used USLE [8].

Until recently, prediction of soil loss rates on National Forest lands involved using the USLE [8, 22]. Soil losses were evaluated in the context of potential soil losses, natural soil losses, current soil losses and tolerable soil losses. Potential losses were those that would occur after complete removal of the vegetation and litter. Natural losses were associated with the potential natural vegetation community. Current losses were those occurring with current management. Tolerable loss was assumed to be the rate that can occur while sustaining inherent site productivity [8].

The Universal Soil Loss Equation (USLE) is a widely used method for calculating annual soil losses, based on rainfall, runoff, slope, runoff length, soil type and landuse parameters. The equation originally developed on small agricultural plots, but has been adopted for evaluating erosion from large watersheds under a wide range of land uses. [41]

$$A = R \times K \times L \times S \times C \times P \quad (1)$$

where A represents the soil loss, commonly expressed in tonnes ha^{-1} year^{-1}. R refers to the rainfall erosivity factor, calculated by the summation of the erosion index EI30 over the period of evaluation. EI30 is a compound function of the kinetic energy of a storm and its 30-min maximum intensity. The latter factor is defined as the greatest average rainfall intensity experiences in any 30-min period during a storm. K is the soil erodibility factor reflecting the susceptibility of a soil type to erosion. It is expressed as the average soil loss per unit of the R factor. L is an index of slope length, S is a slope gradient index, C is an index for the protective coverage of canopy and organic material in direct contact with the ground. It is measured as the ratio of soil loss from land cropped under specific conditions to the corresponding loss from tilled land under clean-tilled continuous fallow conditions. Finally, the protective factor P represents the soil conservation operations or other measures that control the erosion, such as contour farming, terraces, and strip cropping. It is expressed as the ratio of soil loss with a specific support practice to the corresponding loss with up-and-down slope culture [41].

Soil loss rates have been generally estimated in agricultural areas up to now. Various USLE and GIS combinations have been used to estimate soil loss in forest land [25]. But in this kind of studies, soil loss was determined by quantitatively. For example; in study realized in Taiwan estimating watershed erosion using GIS coupled with the USLE in agricultural areas. Furthermore a WinGrid system was developed to calculate slope length factor (L) in USLE [4].

Samar [30] developed three soil loss prediction models (WEPP, EPIC, ANSWERS) and used them for simulating soil loss and testing their capability in predicting soil losses for three tillage systems (rigde-till, chise-plow, and no-till). In other study (leave a space after point), USLE and GIS combination were used to predict long-term soil erosion and sediment transportation from hillslopes to stream networks under different climate conditions and forest management scenarios. Soil erosion was predicted by the USLE watershed level. The GIS utilities are employed to calculate total mass of sediment moving from each cell to nearest stream network [35]. Mısır et al. [25] developed a soil loss model applicable for forest management scenarios for forested areas in northern Turkey.

Forest values including soil protection function need to be determined quantitatively in multi-objective forest management planning. Relationships between soil loss and stand structure on a particular must be determined before incorporation of soil protection values into multi-objective forest management plans.

5. Soil Loss Estimation

The soil loss expressed as ton ha^{-1} year^{-1} is determined using the Universal Soil Loss Equation (USLE). Soil samples are collected from sample plots and analyzed in a laboratory for soil properties including; silt %, sand %, clay %, organic matter %, and classes for structure and permeability. The soil erodibility factor K values of soil samples are calculated using the following equation [41]:

$$K = \frac{2.1 \times M^{1.14} \times 10^{-4} \times (12 - OM) + 3.25 \times (S\text{-}2) + 2.5 \times (P\text{-}3)}{100} \quad (2)$$

where *OM* is soil organic matter content, *M* is (%silt + %very fine sand)x(100-%clay), *S* is soil structure code and *P* is permeability class. If soil organic matter content was greater or equal to 4%, *OM* was considered constant at 4%. Moreover, the influence of rock fragments on soil loss was accounted for by a subsurface component in the soil erodibility *K* factor [29]. The rainfall erosivity was differently obtained from average annual rainfall erosivity map for countries or locations.

The slope length factor L, accounts for increases in runoff volume as downslope runoff lengths increase. The slope stepness factor S accounts for increased runoff velocity as slope stepness increases. These factors were obtained from digitized topographic maps of scale 1:25 000.

For direct application of the USLE a combined slope length and slope stepness (LS) factor was evaluated for each sample plots as [1]:

$$LS = l^{0.5} \times (0.0138 + 0.00965 \times S + 0.00138 \times S^2) \tag{3}$$

where l is runoff length (meter), S is slope (percent).

Crop and management factor is the soil loss from an area with specified cover. C is a function of landuse conditions such as vegetation type, before and after harvesting, crop residues, and crop sequence. Forest management practices create a variety of conditions that influence sheet and rill erosion. The USLE has been used with varying degrees of success to predict these forms of erosion on forest land. Assigning a proper value to cover-management factor (C) in the USLE is a problem, however. An undisturbed, totally covered forest soil usually yields no surface runoff. What erosion does occur on undisturbed forest land comes from stream channels, soil creep, landslide, gullies, and pipes, none of which are evaluated by the USLE. Logging, road building, site preparation, and similar activities that disturb and destroy cover expose the soil to the erosivity of rainfall and runoff [41].

Tree categories of woodland are considered separately:

1. undisturbed forest land,

2. woodland that is grazed, burned, or selectively harvested, and

3. forest lands which have had site preparation treatments for re-establishment after harvest.

Factor C for undisturbed forest land may be obtained from Table 1 [9].

Percent of area covered by canopy of trees	Factor *C*
100 – 75	0.0001 – 0.001
70 – 45	0.002 – 0.004
40 – 20	0.003 – 0.009

Table 1. Factor *C* for undisturbed forest land

The conservation practice factor P, is determined by the extend of conservation practices such as strip, cropping, contouring, and terracing practices, which tend to decrease the erosive capabilities of rainfall and runoff. Values of P range from zero to one.

6. Data Analysis and Modeling

The candidate variables modeling are numerous and diverse. Hartanto et al. [14] classified such variables in four groups: Soil characteristics, physiographic properties, climatic properties and stand characteristics. The candidate variables of soil loss models can be divided in to two groups:

1. measures of physiographic structure and
2. measures of the stand level of structure and density.

Altitude, exposition, aspect, slope and exposure length have been used as measures of physiographic structure. Mean height, mean diameter, crown closure and stand density may have been used as measures of the stand level of structure.

Several possibilities exist to describe stand density. Hamilton [13], Ojansuu et al. [26], Vanclay [38], Thus [37], all of whom used *BA*, and [3], who used *N*, have provided examples of models with stand density parameters as explicatory variables in modeling. Since *N* and *BA* were directly determined, and did not rely on functional relationships, as opposed to volume (*V*), different stand density indexes [7, 28, 10, 5] may be tested.

The soil loss model should be applicable to different stand structures. Therefore, all variables must be tested. Based on the discussion above, the following soil loss models have been generally hypothesized:

$$\hat{A} = \beta_0 + \beta_1 S_1 + \beta_2 S_2 + \beta_3 S_3 \quad (4)$$

where S_1 is the physiographic structure (altitude, exposition, aspect, slope and exposure length), S_2 is the stand structure ($\bar{d}_q$, $\bar{h}_q$ and crown closure) and S_3 is the stand density.

Relationship between magnitude of soil loss obtained from sample plots and stand characteristics have been used to model soil protection value one of the forest values for quantifying soil loss by using linear, nonlinear, mixed linear and mixed nonlinear procedures in Regression Analysis Method The significance of parameter estimates was tested by means of *t*=b/ASE, where *b* is the parameter estimate and *ASE* is the asymptotic standard error. The parameters of the model for data have been determined using a software package (e.g. SPPS, SAS). Only were variables which are significant ($P<0.05$) included in the equation. A soil loss model is constructed based on some site and stand characteristics as a predictor and possible insignificant predictor are excluded. The predicted variable in the soil loss model is annual soil loss amount, which resulted in a linear or nonlinear relationship between the dependent and independent variables. The predictors of a soil loss model were chosen from stand level characteristics as well as their transformations. Some of them had to be significant at the 0.05 level without any systematic errors in residuals. The assumption of homoscedasticity has been tested using the Durbin-Watson test.

7. Model Validation

The soil loss model was evaluated quantitatively by examining the magnitude and distribution of residuals to detect any obvious patterns and systematic discrepancies, and by testing for bias and precision to determine the accuracy at model predictions [39, 33, 11, 20]. Relative bias and root mean square error have been calculated as follows:

$$Bias = \frac{\sum_{i=1}^{n}\left(A_i - \hat{A}_i\right)}{n} \tag{5}$$

$$RMSE = \sqrt{\frac{\sum_{i=1}^{n}\left(A_i - \hat{A}_i\right)^2}{n-p}} \tag{6}$$

where n is the number of observations, p is the number of parameters in the model, A_i and $\hat{A}_i$ are observed and predicted soil loss values, respectively.

In addition, the models were further validated by an independent control data set. The validation of a model should involve independant data. Data were partitioned in two independent groups, one for model development of soil loss estimation and the other set for validation. The data set used for model development of soil loss eestimation comprised approximately 80% of the plots, while the remaining 20% of plots were used for validation. Although the number of sample plots determined for development of soil loss estimation was made relatively large in order to provide sufficient data for model development phase, the number of sample plots in the test data still should be large enough for validation and appropriate statistical test. The deviations between predicted and observed values were tested by Student's Paired-t test or Wilcoxon test.

8. Sample size

The size of sample plot for sampling can be an advantage or disadvantage to model soil loss. A plot size of 800 m^2 means that a relatively large number of the trees are not affected by the forest conditions outside the plot. In other words, a relatively number of trees is affected by the forest conditions inside the plot. In this kind of studies plots that might have been subjected to any harvesting operation between the measurements were excluded from the data material because of insufficient information about treatments. If the harvest on these plots was a result of "regular" management practices, there were no problems related to the exclusion [37]. However, if the harvest was a result of an extraordinary situation (i.e. floods), exclusion of the plots may have lead to an underestimated soil loss amount.

9. Uncertainty

There are many sources of uncertainties related to large scale forestry analyses in general, e.g. related to the inventory of input variables used as basis for the analyses [e.g. 16], to model errors of the numerous functions used for predictions [e.g. 15], to the stochasticity of future condition [e.g. 18, 27] and to the stochasticity of future prices and costs [e.g. 34, 19]. Thus [37], as long as the soil loss models are unbiased, they will not introduce any substantial change with respect to the final uncertainty of large scale forestry analyses.

10. Conclusions

Soil loss is an important variable which is used for multiple forest management planning.

Measuring soil loss is costly; however, foresters usually welcome an opportunity to estimate this function (forest value) with an acceptable accuracy. Missing soil losses may be estimated using a suitable soil loss equation. Based on a comprehensive data set which includes very different stands, such soil loss equation should be fitted for a major tree species in complex.

The stand position and stand density measures used in this kind of studies and variables entered to the soil loss model are easily obtained and are available in forest inventories. In summary, the suggested or developed soil loss models improve the accuracy of soil loss prediction, ensure compatibility among the various estimates in a forest management scenario, and maintain projections with reasonable biological limits.

Linear, nonlinear or mixed models for prediction of soil loss for stand level, designed for use in large scale forestry scenario models and analyses, may been developed. Although soil loss as a phenomenon is complicated to model, and in spite of several uncertain topics revealed from the work, the model fit and the validation tests may be turned out satisfactory.

Provided the many uncertainties of large scale forestry scenario analyses in general, soil loss models seem to hold an appropriate level of reliability, and we feel that it can be applied in such analyses. This does not mean that the model cannot be enhanced, however. With new rotations of permanent sample plots measurements, the models should be evaluated and, if necessary, revised or calibrated.

Author details

Nuray Misir[1*] and Mehmet Misir[2]

*Address all correspondence to: nuray@ktu.edu.tr

1 Karadeniz Technical University, Faculty of Forestry, Turkey

2 Department of Forest Management, Turkey

References

[1] Arnoldus, H. M. J. (1977). Predicting soil losses due to sheet and rill erosion. *FAO conservation guide no.1,Guidelines for watershed management, Rome, Italy.*

[2] Asan, Ü. (1992). letme sınıfı ayrımında fonksiyonel yaklaşım. *Orman Mühendisliği Dergisi*, 5, 30-31.

[3] Burgman, M., Incoll, W., Ades, P., Ferguson, I., Fletcher, T., & Wholers, A. (1994). Mortality models for mountain and Alpine Ash. *Forest Ecology and Management*, 67, 319-327.

[4] Chao-Yuan, L., Wen-Tzu, L., & Wen-Chieh, C. (2002). Soil erosion prediction and sediment yield estimation: the Taiwan experience. *Soil & Tillage Research*, 68, 143-152.

[5] Chisman, H. H., & Schumacher, A. F. (1940). On the tree-area ratio and certain of its applications. *Journal of Forestry*, 38, 311-317.

[6] Clarke, A. C. (1998). President, Experts and Asteroids. *Science*, June, 5, 1532-1533.

[7] Curtis, R. O., Clendenan, G. W., & Demars, D. J. (1981). A new stand simulator for coast douglas-fir: DFSIM users guide: U.S. forest service general technical report PNW-128.

[8] De Bano, L. F., & Wood, M. K. (1990). Soil loss tolerance as related to rangeland productivity. Proceedings of the Soil Quality Standards Symposium. U.S. Department of Agriculture, Forest, Service Publication WO-WSA-2. , 15-27.

[9] Dismeyer, G. E., & Foster, G. R. (1984). Estimating the cover-management factor (C) in the Universal Soil Loss Equation for forest conditions. *Journal of Soil and Water Conservation*, 36, 235-240.

[10] Drew, T. J., & Flewelling, J. W. (1977). Stand density management: an alternative approach and its application to Douglas-Fir plantations. *Forest Science*, 25, 518-532.

[11] Gadow, K., & Hui, G. (1998). Modelling forest development. *Faculty of forest and woodland ecology, University of Göttingen.*

[12] Grigal, D. F. (2000). Effects of Extensive Forest Management on Soil Productivity. *Forest Ecology and Management*, 138(1-3), 167-185.

[13] Hamilton, D. A. (1986). A Logistic model of mortality in thinned and unthinned mixed conifer stands of Northern Idaho. *Forest Science*, 32, 989-1000.

[14] Hartanto, H., Prabhu, R., Widayat, A. S. E., & Adsak, C. (2003). Factors affecting runoff and soil erosion: Plot-level soil loss monitoring for assessing sustainability of forest management. *Forest Ecology and Management*, 141, 1-14.

[15] Kangas, A. (1996). On the bias and variance in tree volume predictions due to model and measurement errors. *Scandinavian Journal of Forest Research*, 11, 281-290.

[16] Kangas, A., & Kangas, J. (1999). Optimization bias in forest management planning solutions due to errors in forest variables. *Silva Fennica*, 33, 303-315.

[17] Kozlowski, T. T. (2000). Responses of woody plants to human-induced environmental stresses : Issues, problems, and strategies for alleviating stress. *Critical Reviews in Plant Sciences*, 19(2), 91-170.

[18] Larsson, M. . (1994). The significance of data quality in compartmental forest registers in estimating growth and non-optimal losses-a study of final fellling Compartments in Northern Sweden. Report 26. Swedish University of Agricultural Sciences Umea, Sweden

[19] Leskinen, P., & Kangas, J. (1998). Analysing uncertainties of interval judgment data in multiple criteria evaluation of forest plans. *Silva Fennica*, 32, 363-372.

[20] Mabvurira, D., & Miina, J. (2002). Individual-tree growth and mortality models for Eucalyptus grandis (Hill) maiden plantations in Zimbabwe. *Forest Ecology and Management*, 161(1-3), 231-245.

[21] Megahan, W. F. (1990). Erosion and site productivity in Western-Montane forestecosystems. Symposium on Management and Productivity of Western-Montane. *Boise, ID, USDA Forest Service General Technical Report INT-280*, 146-150.

[22] Megahan, W. F. (1992). Logging erosion sedimentation… Are they dirty words? *Journal of Forestry*, 70(7), 403-407.

[23] Mısır, M. (2001). Developing a Multi objective model forest management plan using GIS and Goal Programming (A case study of Ormanüstü Planning Unit). *Karadeniz Technical University, PhD Thesis, Trabzon.*

[24] Mısır, M., & Başkent, E. Z. (2002). The role of GIS in Multi objective forest planning. *International Symposium on GIS, İstanbul, Tukey, Proceedings*, 449-465.

[25] Mısır, N., Mısır, M., Karahalil, U., & Yavuz, H. (2007). Characterization of soil erosion and its implication to forest management. *Journal of Environmental Biology*, 28(2), 185-191.

[26] Ojansuu, R., Hynynene, J. , Koivunen, J., & Luoma, P. Luonnonprosessit metsalaskelmassa. 1991, *METSA 2000*, 385, 1-59.

[27] Pasanen, K. (1998). Integrating variation in tree growth in to Forest Planning. *Silva Fennica*, 32, 11-25.

[28] Reineke, L. H. (1933). Perfecting a stand density index for even-aged forests. *Journal of Agricultural Research*, 46(7), 627-638.

[29] Renard, K. G., Foster, G. R., Weesies, G. A., Mc Cool, D. K., & Yoder, D. C. (1997). Predicting soil erosion by water: A guide to conservation planning with the Revised

Universal Soil Loss Equation (RUSLE). *USDA Agricultural Research Service Handbook* [703].

[30] Samar, J., Bhuyan, Prasanta. K., Kalita, Keith. , & Janssen, A. (2002). Soil loss predictions with three erosion simulation models. *Environmental Modeling*, 17(2), 137-146.

[31] Schmidt, B. L., Allmaras, R. R., & Mannering, J. V. (1982). Preface in: Determinants of soil loss tolerance. *ASA Special Publication,No. 45, Am. Soc. Agr., Madison, Wiscon.*

[32] Schumm, S. A., & Harvey, M. D. (1982). Natural Erosion in USA. *Schmidt, B.L., ed. Determinants of Soil Loss Tolerance. Madison, WI: American Society of Agronomy, Special Publication*, 45, 23-29.

[33] Soares, P., Tomé, M., Skovsgaard, J. P., & Vanclay, J. K. (1995). Evaluating a growth model for forest management using continuous forest inventory data. *Forest Ecology and Management*, 71, 251-265.

[34] Stähl, G. (1994). Optimal stand level inventory intensities under deterministic and stochastic stumpage value assumptions. *Scandinavian Journal of Forest Research*, 9, 405-412.

[35] Sun, G., & Mc Nulty, S. D. (1998). Modeling soil erosion and transport on forest landscape. *Conference 29, NV. Steamboat Springs, Co: International Erosion Control Association*, 187-198.

[36] Swanson, F. J., Clayton, J. L., Megahan, W. F., & Bush, G. (1989). Erosional processes and long-term site productivity. *Perry, D. A., Meurisse, R.; Thomas, B., Miller, R., Boyle, J., Means, J., Perry, C. R., Powers, R. F., eds. Maintaining The Long-Term Productivity of Pacific Northwest Forest Ecosystems. Portland, OR: Timber Press*, 67-81.

[37] Thus, E. (1997). Naturlig avgang av traer (Natural mortality of trees). *Raport Fra Skogfrosk*, 6.

[38] Vanclay, J. K. (1991). Mortality functions for North Queensland rain forests. *Journal of Tropical Forest Science*, 4(1), 15-36.

[39] Vanclay, J. K. (1994). Modelling forest growth and yield, applications to mixed tropical forests,. *CAB International, Wallingford, UK.*

[40] Weetman, G. F. (1998). A forest management perspective on sustained site productivity. *Forestry Chronical*, 74, 75-77.

[41] Wischmeier, W. H., & Smith, D. D. (1978). Predicting rainfall erosion losses-a guide to conservation planning . *USDA agricultural research service handbook, no. 537, Washington, D.C.*

Chapter 6

Change of Soil Surface Roughness of Splash Erosion Process

Zicheng Zheng and Shuqin He

Additional information is available at the end of the chapter

http://dx.doi.org/10.5772/51278

1. Introduction

Soil erosion is a common global environmental problem and undermines sustainable development in various economies and societies. Detailed information about changes in surface roughness during the whole soil erosion process remains limited, however, due to practical difficulties in obtaining direct soil microrelief measurements (Huang, 1998) and a lack in systematic research. The Chinese Loess Plateau is one of the most severely eroded regions in the world, which has created many environmental problems along the lower reaches of the Yellow River. Despite this, however, very little erosion-based research has been conducted on the Loess Plateau. Erosion and runoff processes are influenced mainly by soil surface characteristics such as soil surface roughness, cohesion, and granular stability. Among these characteristics, soil surface roughness is a key parameter (Gómez, and Nearing, 2005; Mirzaei et al., 2008), and is used to describe the variation in surface elevation across a field. The soil surface micro-topography or roughness is strongly influenced by agricultural activities, together with soil properties and climate. The term soil roughness was used to describe disturbances or irregularities in the soil surface at a scale which was generally too small to be captured by a conventional topographic map or survey. Soil surface roughness is an important parameter in understanding the mechanisms of soil erosion by water and wind. Many erosion related surface processes, such as depression water storage, raindrop or wind shear detachment, and sediment transport have characteristic lengths in millimeter scales. Thus, soil surface roughness resulting from small scale elements is important in understanding these processes and their spatial variation (Huang and Bradford, 1990). Soil surface roughness determines the storage of water on the soil surface and may indirectly influence its infiltration capacity. The velocity of overland flow is controlled by the hydraulic resistance of the soil surface. Soil surface roughness affects the organization of the drainage pattern on the field and the catchments scale, which in turn may have important implications for the spatial distribution of sediment sources and sinks. Conversely, some of these processes affect surface roughness. Most of the literature on soil surface roughness focusing on its mathematical description and on how it changes under rainfall(Linden and Van doren,1986; Römkens and Wang,1987; Lehrsch et al,1988; Bertuzzi et al.,1990).Soil surface roughness significantly impacts runoff and sediment generation under rainfall in several different ways.

It was one kind of erosion phenomenon which the raindrop strikes the soil surface to create the soil particle dispersion and the leap moves for the splash erosion. It was one of the important components to soil erosion (Wang et al,1997, 1999; Zhao and Wu,2001; Liu and Wu,1996;Wu, 1999; Wu and Zhou,1994). The kinetic energy which the raindrop dropped from airborne was the higher than that of sheet flow and erosion sediment during the rainfall runoff for the different soil surface (Huang,1983). According to the observation data of some researches, the soils of bare land by the raindrop scattered were 10 times than those of the laminar flow scoured (Cai et al,1998). Many authors have studied the effect of rainfall on soil surface roughness and developed models to describe the change of soil surface roughness. Some researches obtained the simple forecast model of soil surface roughness (Johson et al,1979; Onstad,1984; Steichen, 1984). Later, the widely accepted concept of decreasing roughness with increasing amount of rainfall or rainfall energy may not always be appropriate. After 63 mm of rainfall the surface was crusted and surface roughness was decreased. However, an additional 92 mm of rainfall appeared to have a higher roughness value (Huang and Bradford,1992).

The objective of this study was to focus on the relationship between soil surface roughness and splash erosion. First, soil surface roughness affected on splash erosion under the condition of rainfall. Second, how was the change of soil surface roughness during the period of rainfall?

2. Material and Methods

2.1. Soil and soil box design

Experiments were carried out at the Northwest AF University Soil Erosion Research Laboratory,Yangling town,China. The soil was collected from the topsoil soil (0-20cm) in Yangling town. Basic properties of soil were following (Table 1).

Particle size/ (%)					
>0.25mm	0.25—0.05mm	0.05—0.01mm	0.01—0.005mm	0.005—0.001mm	<0.001mm
0.12	2.70	41.13	6.88	12.89	36.28

Table 1. Particle size distribution (0—20cm) of experimental soil.

Four iron boxes of 2.0 m×1.0 m×0.5 m were used in the rainfall simulation study. Air-dried top soil was passed through a 10mm sieve to insure homogeneity and placed in every erosion box with an area of $2m^2$. The soil bulk density was controlled to 1.08 g cm^{-3} in order to assure to fill to be homogeneous and close natural state through randomization method. Before the rainfall, the soil mechanical composition was measured by the pipette method, and the soil bulk density was measured by the ring sampler method.

2.2. Rainfall simulations and soil surface roughness measurement techniques

Rainfall/erosion methods in this lab study were similar to those described by Zheng et al. (2007). The soil box was adjusted at 150 slope gradient and then placed under a rainfall simulator with oscillating nozzles. Rainfall high was 2.7 m and effective rainfall area approximately was 20 m^2. This experiment used the constant rainfall intensity, therefore, different rainfall intensities were rated before testing. The uniformity of rainfall was up to 0.90. Development of micro-relief was monitored by recording soil surface at the beginning and at the end of the experiments, using the non-contact profile laser scanner measuring instrument specified and calculated (Zheng,2007). The maximum range of detectable elevation differences was approximately 500 mm. Surface relief was measured point by point in a regularly spaced grid. The maximum scanning area was 2 m. The surface roughness was measured for each soil box before the rainfall and after the rainfall separately with non-contact the profile laser scanner.

Simulated rainfall for each replication of a treatment were divided into the single rainfall intensity and the combined rainfall intensity, the parameters of single rainfall intensity respectively were 0.68 mm/min and 1.50 mm/min, the parameter of combined rainfall intensity is 0.68 mm/min,1.00 mm/min and 1.50 mm/min. The above experiments had three repeats. Each experiment started on a freshly prepared surface for each replication of a treatment. The rainfall simulation duration were depended on the change of soil surface.

2.3. Management treatments

The four artificial management measures were designed according to the local agriculture custom in Loess Plateau, because agriculture management measures were mainly artificial management. The four artificial management measures were the raking cropland (PM), the artificial hoe (CH), the artificial dig (TW) and the contour slope (DG).They were used to simulate different types of soil surface roughness separately, the straight slope (CK) was taken to the control.

2.4. Splash erosion

The amounts of splash erosion were collected through the hanging splash erosion board and measured by the oven drying method. The width to the hanging splash erosion board was 1m and the height was 0.5m.The hanging splash erosion board was installed in the middle of the soil box was to be used to collect splashing soil during the experiments(Fig.1).

At the same time, raindrops of every rainfall were collected to calculate raindrop diameter. Raindrop diameter was measured through the color spot method according the B.Z.Dou et.al (Dou and Zhou,1982; Zheng and Gao,2000), the formula was as following:

$$d = 0.356 D^{0.712} \quad (1)$$

where d is the raindrop diameter of every rainfall (mm), D is the color spot diameter (mm).

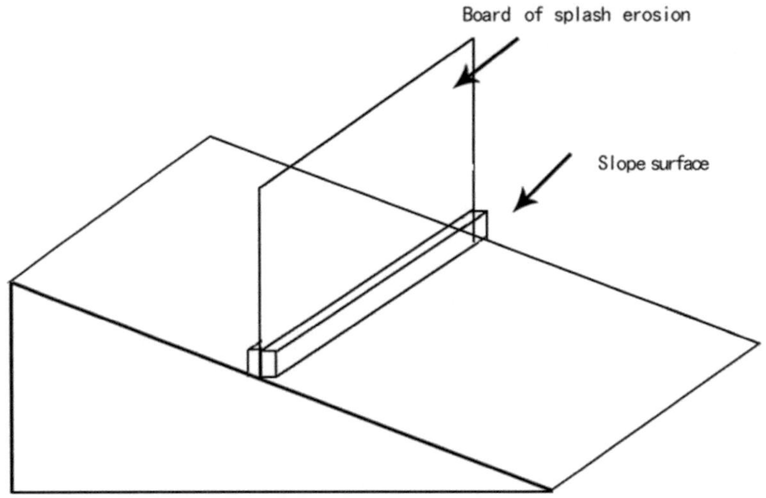

Figure 1. Collecting board of splash erosion.

3. Results and Discussion

3.1. Changing characteristics of soil surface roughness on the single rainfall intensity

The changing characteristics of soil surface roughness had complicated relatively under the different rainfall intensity (Table.2). The soil surface roughness increased on the CK slope under the rainfall intensity of 0.68 mm/min. The soil surface roughness decreased on the CK slope under the rainfall intensity of 1.50 mm/min. On the PM slope, the changing characteristics of soil surface roughness was consistent with the CK slope, however, changing characteristics of soil surface roughness decreased on the other slopes under the rainfall intensity of 0.68 mm/min. Under the rainfall intensity of 1.50 mm/min, the changing characteristics of soil surface roughness with other slopes were contrary with the CK and showed increasing trends.

The reasons of the above results were the interaction among raindrop kinetic energy, soil surface roughness and splash amount possibly. From the angle of physics, the function of raindrop to the soil surface was one kind of acting process actually. The raindrop would hit and compact exposed soil surface when the rainfall began. At the same time, infiltrate ability of the soil reduced and soil bulk density increased gradually, and the partial soil was easy to form the crust due to soil surface fine-grain inserting in former place or migration and jamming soil pore space. Thus, soil surface roughness and the splash amounts also changed.

Tillage practice	Rainfall intensity/ mm·min^{-1}	R_0/cm	R/cm	R/R_0
Straight slope (CK)	0.68	0.201	0.235	1.169
	1.50		0.193	0.960
Raking cropland (PM)	0.68	0.240	0.246	1.024
	1.50		0.268	1.115
Artificial hoe (CH)	0.68	0.706	0.654	0.926
	1.50		0.732	1.036
Artificial dig (TW)	0.68	0.812	0.701	0.864
	1.50		0.874	1.077
Contour slope (DG)	0.68	1.633	1.576	0.965
	1.50		1.707	1.045

Note: R_0-soil surface roughness before rainfall, cm*R*- soil surface roughness after rainfall, cm. The same bellow.

Table 2. Change of soil surface roughness on the single rainfall intensity.

Relationships between rainfall energy and soil surface roughness were obtained by the method of statistics and analysis. The results followed:

Under the rainfall intensity of 0.68 mm/min: R_1/R_0=49261E$^{-3.3451}$ r=0.817 n=15

Under the rainfall intensity of 1.50 mm/min: R_1/R_0=2×10^{6}E$^{-4.2309}$ r=0.836 n=15

where R_1 is the soil surface roughness after rainfall(cm), R_0 is the soil surface roughness before rainfall(cm), E is the total kinetic energy of raindrop (J/cm^2 min), n is the sample number.

They had the power function relationship between the change of the soil surface roughness and kinetic energy of raindrop under the different rainfall intensities. Soil surface roughness decreased with the increasing kinetic energy of raindrop. The results had the consistent with Burwell (1969) and Steichen (1984).

3.2. Changing characteristics of soil surface roughness under the combined rainfall intensity

The combined rainfall intensity was be simulated in order to clear about the change and nature of soil surface roughness. The changing characteristics of soil surface roughness were different for the different slopes under the combined rainfall intensity (Table.3). The changing characteristics of soil surface roughness increased first, and then decreased, and increased finally with the increasing rainfall intensity on the CK slope. However, the changing characteristics of soil surface roughness increased on the PM slope, and the change of soil surface roughness increased first and then decreased on other slopes with the increasing rainfall intensity.

Tillage practice	Rainfall intensity/mm·min^{-1}	R_0/cm	R/cm
Control slope (CK)	0.68	0.201	0.235
	1.00		0.222
	1.50		0.302
Raking cropland (PM)	0.68	0.240	0.246
	1.00		0.283
	1.50		0.319
Artificial hoe (CH)	0.68	0.706	0.654
	1.00		0.624
	1.50		0.625
Artificial dig (TW)	0.68	0.812	0.701
	1.00		0.586
	1.50		0.614
Contour slope (DG)	0.68	1.633	1.576
	1.00		1.572
	1.50		1.577

Table 3. Change of soil surface roughness under the combined rainfall intensity.

The reasons of the above results were the interaction between raindrop kinetic energy and soil surface roughness. The micro-relief of CK slope and PM slope were relatively small in the initial period of the rainfall. At the same time, the raindrop impact was relatively even, and they had the positive relationship between the raindrop kinetic energy and the rainfall intensity. Therefore, the changing characteristics of soil surface roughness increased and the raindrop impact gradually strengthened with the increasing rainfall intensity for the CK slope and PM slope. However, the micro-relieves of other slopes were relatively obvious in the initial period of the rainfall. At the same time, the convex fraction of raindrop impact was splashed and the concave fraction of raindrop impact was padded by other soil particle, and the part of the concave appeared the crust. So, the soil surface roughness decreased in the initial period of the rainfall. The partial soil particle of surface was dispersed or migrated, caused soil surface roughness to increase with the continuous the function of raindrop impact.

The changing characteristics of soil surface roughness were decided on the initial soil surface condition and the surface dynamic process of rainfall. The changing characteristics of soil surface roughness were analyzed with the impact of accumulating rainfall amount under the combined rainfall intensity in order to clarify the change of soil surface roughness. The results followed as Fig. 2.

Relationships between the accumulated rainfall amount and the change of soil surface roughness were obtained by the method of statistics and analysis.

$$R/R_0 = -6\times10^{-6}P^3 + 0.0012P^2 - 0.0768P + 2.3629 \quad r = 0.708; n = 15 \qquad (2)$$

where R_1 is the soil surface roughness after rainfall(m), R_0 is the soil surface roughness before rainfall (m), P is the accumulated rainfall amount(mm), n is the sample number.

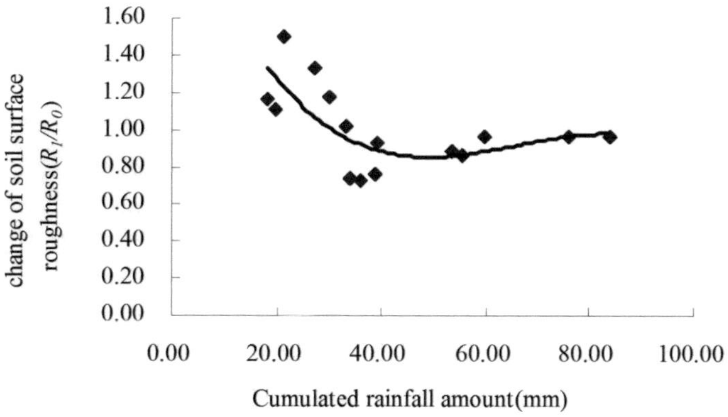

Figure 2. Relationship between change of soil surface roughness and cumulated rainfall amount.

The changing characteristics of soil surface roughness increased first and then decreased for all the slops with the increasing accumulated rainfall amount.

3.3. Relationship between the changing of soil roughness and splash erosion amount

Table.4 shows the splash erosion amounts of all tillage practices under the different rainfall intensities.

Tillage practice	Rainfall intensity/ $mm\cdot min^{-1}$	Splash erosion amounts / $g\cdot cm^{-2}$
Control slope (CK)	0.68	0.10
	1.50	0.57
Raking cropland (PM)	0.68	0.93
	1.50	1.18
Artificial hoe (CH)	0.68	1.27
	1.50	1.46
Artificial dig (TW)	0.68	0.61
	1.50	1.09
Contour slope (DG)	0.68	0.81
	1.50	1.01

Table 4. Splash erosion amounts under the different rainfall intensities.

The change of splash erosion amounts had the difference under the different rainfall conditions on the all the slopes. The splash erosion amounts of the CK slope were lower than those of other slopes under the rainfall intensity of 0.68 mm/min and 1.50 mm/min (Table.4).

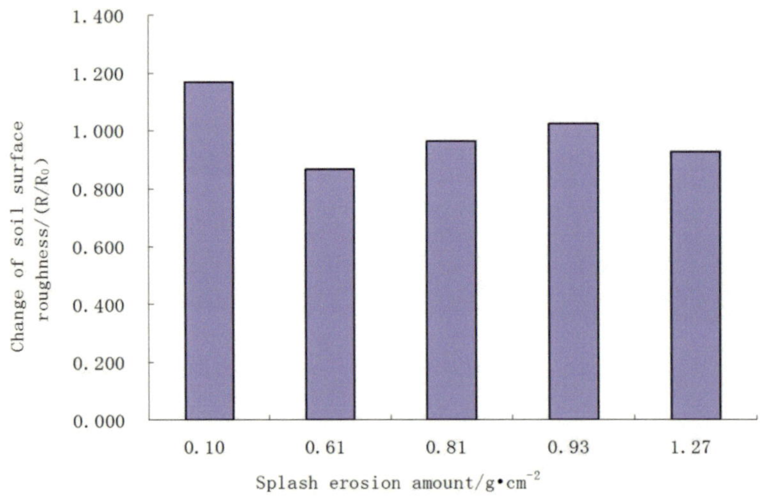

(a) I=0.68 mm/min

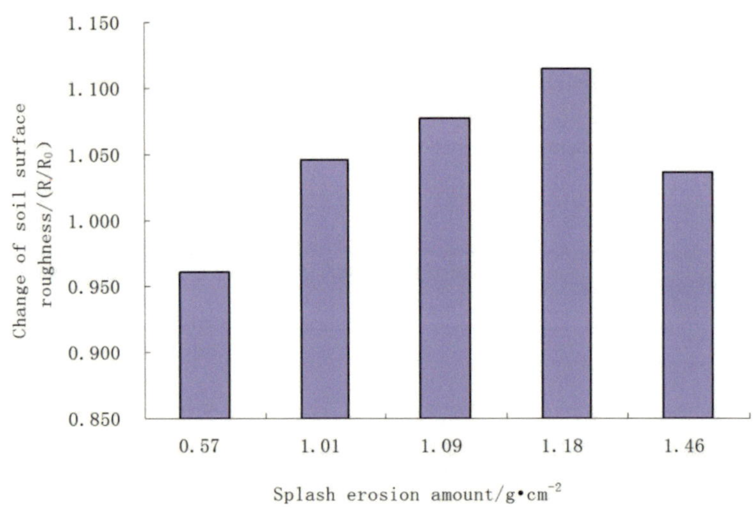

(b) I=1.50 mm/min

Figure 3. Relationship between soil surface roughness and splash erosion amounts under the different rainfall intensities.

The change of soil surface roughness showed the different characteristic with the splash erosion amounts under the different rainfall conditions for the all the tillage practices. The splash erosion amounts of the CK slope were lower than those of other slopes under the rainfall intensity of 0.68 mm/min, but the change of soil surface roughness was the highest (Fig.3a). However, the splash erosion amounts of the CK slope were lower than those of other slopes under the rainfall intensity of 1.50 mm/min, and the change of soil surface roughness was the lowest (Fig.3b).

The change of soil surface roughness increased first and then decreased for other slopes with the increasing splash erosion amounts under the rainfall intensity of 0.68mm/min and 1.53 mm/min. The above results were caused the interaction of raindrop kinetic energy and soil surface fluctuation condition. The raindrop impact to rainfall intensity of 1.50 mm/min was obviously stronger than that of the rainfall intensity of 0.68 mm/min, and soil particles of former sites were sputtered. In turn, the around particle of the former sites might supplied soil particles through the same action. The soil particles of the continuous supplement might also supply the material base for the migration. The unceasing replacement would cause the interaction of soil surface roughness and splash erosion. So, the results were quite complicated.

4. Conclusions

Under the rainfall intensity of 0.68 mm/min, the soil surface roughness increased on the control slope, the changing characteristics of soil surface roughness to the raking cropland slope was consistent with the control slope, however change of soil surface roughness to the other slopes decreased. The splash erosion amounts of the control slope were lower than those of other slopes, but the change of soil surface roughness was the highest. Under the rainfall intensity of 1.50 mm/min, the soil surface roughness decreased on the control slope, the change of soil surface roughness showed increasing trends on the other slopes. The splash erosion amounts of the control slope were lower than those of other slopes, and the change of soil surface roughness was the lowest. Under the combined rainfall intensity, the change of soil surface roughness of the control slope increased first, and then decreased, and increased finally with the increasing rainfall intensity. The change of soil surface roughness increased on the raking cropland slope, and the change of soil surface roughness increased first and then decreased for other slopes with the increasing rainfall intensity.

Acknowledgements

The research was supported by the National Natural Science Foundation of China (Grant No. 40901138),National Basic Research Program of ChinaGrant No. 2007CB407201and also supported by State Key Laboratory of Soil Erosion and Dryland Farming on the Loess PlateauInstitute of Water and Soil ConservationChinese Academy of Sciences and Ministry of Water Resources (Grant No. 10501-283).

Author details

Zicheng Zheng[1,2*] and Shuqin He[1,2]

*Address all correspondence to: zichengzheng@yahoo.com.cn

1 College of Resource and Environment, Sichuan Agriculture University

2 State Key Laboratory of Soil Erosion and Dryland Farming on the Loess Plateau Institute of Water and Soil Conservation Chinese Academy of Sciences and Ministry of Water Resources, China

References

[1] Bertuzzi, R., Rouws, G., & Couroult, D. (1990). Testing roughness indices to estimate soil surface roughness changes due to simulated rainfall. *Soil Tillage Res.*, 17, 87-99.

[2] Burwell, R. E., & Larson, W. E. (1969). Infiltration as influenced by tillage-induced random roughness and pore space. *Soil Sci. Soc. Am. Proc.*, 33, 449-452.

[3] Cai, Q. G., Wang, G. G., & Chen, Y. Z. (1998). *Process and simulate of sediment in small watershed of Loess Plateau*, Beijing, Science Press.

[4] Dou, B. Z., & Zhou, P. H. (1982). Method of measure and mathematics to raindrop. *Bulletin of Soil and Water Conservation*, 2(1), 44-47.

[5] Go′mez, J. A., & Nearing, M. A. (2005). Runoff and sediment losses from rough and smooth soil surfaces in a laboratory experiment. *Catena*, 59, 253-266.

[6] Huang, C. H. (1998). Quantification of soil microtopography and surface roughness. Fractals in soil scienceIn: Baveye, P., Parlange, J.Y., Eds. B.A.Stewart. *Advances in Soil Science*, CRC.

[7] Huang, Bingwei. (1983a). Problems of soil conservation in the middle of Yellow River. *Soil and Water Conservation in China* [1], 8-13.

[8] Huang, C. H., & Bradford, J. M. (1990). Depressional storge for Markor-Gaussian surfaces. *Water Resources Research*, 26, 2235-2242.

[9] Huang, C. H., & Bradford, J. M. (1992). Applications of a laster scanner to quantify soil microtopogrophy. *Soil Sci. Soc. Am. J*, 56(1), 14-21.

[10] Johnson, C. B., Mannering, J. V., & Moldenhauer, W. C. (1979). Influence of surface roughness and clod size and stability on soil and water losses. *Soil Sci. Soc. Am. J*, 43, 772-777.

[11] Lehrsch, G. A., Whisler, F. D., & Romkens, M. J. M. (1987). Soil surface roughness as influenced by selected soil physical properties. *Soil Tillage Res.*, 10, 197-212.

[12] Linden, D. K., Van , D. M., & Doren, J. R. (1986). Parameter for characterizing tillage-induced soil surface roughness. *Soil. Sci. Soc. Am. J*, 50, 1561-1565.

[13] Liu, Bingzheng., & Wu, Faqi. (1996). *Soil Erosion*, Xi'an, People Press of Shannxi.

[14] Mirzaei, M. R., Ruy, S., Ziarati, T., Khaledian, M. R., & Christina, B. (2008). Changes of soil surface roughness under simulated rainfall evaluated by photogrammetry. *Geophysical Research Abstracts*, 10, EGU2008A-02576.

[15] Onstad, C. A. (1984a). Depressional storage on tilled soil surfaces. *Trans. Am. Soc. Agric. Eng.*, 27, 729-732.

[16] Romkens, M. J. M., & Wang, J. Y. (1987). Soil roughness changes of tillage system from rainfall. *Trans. Am. Soc.of Agric. Eng.*, 30, 101-107.

[17] Steichen, J. M. (1984). Infiltration and random roughness of a tilled and untilled clay pan soil. *Tillage Res*, 4, 251-262.

[18] Wang, X. K., Ao, R. Z., & Yu, G. L. (1999). Rainsplash erosion and its sediment transport capacity on slope. *Journal of Sichuan University*, 3(2), 7-12.

[19] Wang, X., & Fang, D. (1997). A physically-based model of rainsplash erosion on the slope. *Journal of Sichuan University*, 1(3), 90-102.

[20] Wu, F. Q. (1999). *Analysis and study on erosion rainfall and productivity of gently sloping farmland*, Yangling, Institute of Soil and Water Conservation.

[21] Wu, P. T., & Zhou, P. H. (1994). The Effects of Raindrop Splash on the Sheet FlowHydraulic Friction Factor. *Journal of Soil Water Conservation*, 8(2), 40-42.

[22] Zhao, X. G., & Wu, F. Q. (2001). Single raindrop splash law and its selection role on soil particles splashed. *Journal of Soil Water Conservation*, 15(1), 43-45, in Chinese.

[23] Zheng, F. L., & Gao, X. T. (2000). *Process and simulate of soil erosion of Loess slope*, Xi'an, People Press of Shannxi.

[24] Zheng, Z. C. (2007). *Study on the effect and change characteristic of soil surface roughness during the course of water erosion*, Yangling, Northwest Agricultural and Forestry University.

Chapter 7

Quantifying Nutrient Losses with Different Sediment Fractions Under Four Tillage Systems and Granitic Sandy Soils of Zimbabwe

Adelaide Munodawafa

Additional information is available at the end of the chapter

http://dx.doi.org/10.5772/3402

1. Introduction

In soil erosion studies too much emphasis has been placed on the weight of soil loss (t/ha), while the real issue is not only about the amount of soil lost or the area of land degraded, but the effect of soil erosion on the productivity of the land. Soil erosion is rated as one of the major threats of sustainable land management, but the research data on the impact of erosion on soil properties and its effect on crop yield is grossly missing (Hudson, 1993), especially in tropical Africa (Kaihura, *et.al.*, 1998). While the process of erosion is somewhat better understood, the resultant changes in the soil properties, the decline in yield and evaluating the loss in productivity should be of concern to the researchers in this region.

On arable land, soil erosion is initiated through tillage. Tillage is the mechanical manipulation of soil for any purpose (Gill and Vanden Berg, 1967). It is an important part of the overall farming system. The primary objectives of tillage, as given by Godwin (1990) and Lobb (1995) are to prepare a desirable seedbed, to control weeds, enhance soil and water storage and retention, manage crop residues and reduce erosion. Tillage can however, either conserve or damage the soil depending on the intensity of inversion and the degree of exposure of the soil to weather conditions. The intensity of soil inversion also influences surface roughness, which in turn determines the sealing tendency of uncovered soil. The rougher the surface, the smaller the raindrop density per unit time and the lower the tendency to seal (Frede and Gaeth, 1995).

Conventional tillage or ploughing promotes soil organic matter loss through disruption of soil aggregates and increased aeration (Angers, N'dayegamiye and Cote, 1993; Beare, Hendrix and Coleman, 1994; Reicosky, *et al.*, 1996; Salinas-Garcia, Hons and Matocha, 1997). Also through ploughing, the crop residues are buried there-by enhancing organic matter decomposition and transformations. Where ploughing is practiced, it is practically impossible to increase organic matter content, even when huge amounts of fertilizer are applied. Reduced tillage intensity on the other hand can result in the maintenance/ increase of more labile fractions of soil organic matter (Angers, N'dayegamiye and Cote, 1993). Combining reduced tillage with surface crop residues not only inhibits the loss of soil organic matter but also improves soil aggregation.

In Zimbabwe soil tillage can be divided into three broad categories namely: Conventional Tillage, Reduced Tillage and Strip Tillage (Willcocks and Twomlow, 1992). Ploughing with a single furrow ox-drawn mould-board plough (conventional tillage) is the most widely used tillage practice in the communal areas of Zimbabwe and is estimated to be practiced on 73 - 90% of the cultivated area. The remainder of the land is ploughed using hired tractor (5 - 25%) and by hand (1 - 15%). Less than 1% is under tillage systems, which conserve soil, moisture, nutrients and/or energy inputs (Working Document, 1990). Reduced tillage involves mainly tied ridging, ripping and hand-hoeing. The tied ridging system is a useful compromise between drainage and storage (Hudson, 1992). Rainwater is retained in the basins to soak into the soil, so very little run-off occurs (Elwell, 1986). The hand-hoeing system is labour intensive and practiced mainly in areas infested with tsetse flies or in cases of extreme lack of draft power (Working document, 1990). Under this treatment, the ground usually has poor cover, the soil tends to compact and no significant soil conservation potential over conventional tillage has been observed (Vogel, 1992).

The ripping system saves on draft power as only the crop rows are opened and no tillage takes place between the crop rows. This means that the timeliness of operations is improved and yields may be improved as according to Oliver and Norton, (1988), low yields in the communal areas are also largely a result of late ploughing/ planting. Two types of ripping systems are currently under research in Zimbabwe, namely ripping into residues and clean ripping, where all crop residues are removed after crop harvest. Clean ripping reduces tillage and draft power requirement, however, the soil and water conservation potential of this system is low. Vogel, (1992) found no significant differences between this system and conventional tillage in terms of run-off and soil loss.

Mulch ripping has a lot of potential in conserving soil and water. Mulching has not yet been promoted in the communal areas as most of the stover is fed to cattle; however, the advantages of the system have been observed. When mulch is left on the soil surface, the soil is protected from high intensity raindrops (Adams, 1966; Elwell, 1986). Run-off, soil loss and subsequent nutrient loss are reduced (Elwell, 1986; Reicosky*et al.*, 1996). The underlying soil retains its high infiltration rate and most of this infiltrated moisture is protected from evaporation (Adams, 1966). The disadvantages are mainly weeds and the carryover of pests and diseases (Braithwaite, 1976; Elwell, 1986).

The effects of conventional tillage on the soil are generally known and can be summed up in a cause/effect relationship as shown in Figure 1. The conservation tillage systems ideally have to be designed in such a way that they reduce the effects of conventional tillage by generally protecting the land and sustaining crop production. Figure 1 tries to summarize the effect of tillage on soil productivity, giving the desired effect should conservation tillage systems be used.

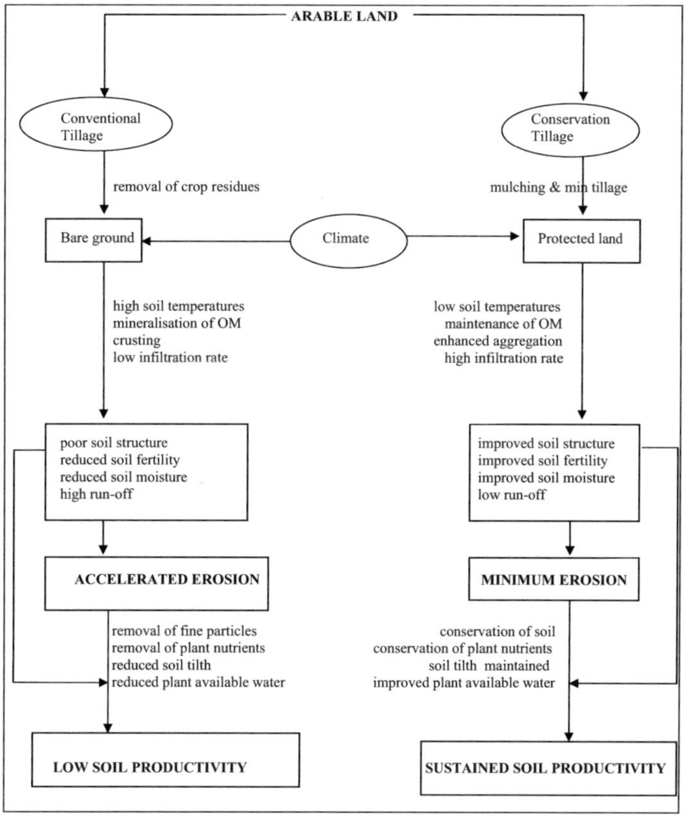

Figure 1. Soil erosion as affected by tillage and climate and its impact on soil productivity

Appropriate tillage systems should therefore, aim to maintain/ increase soil organic matter as it is the key to the productivity of the soils, as will be highlighted shortly. Organic matter content in most agricultural soils has been found to be highly correlated with their tilth, fertility and potential productivity. This soil constituent has positive effects on soil chemical, physical and biological properties that in turn contribute to improved crop yields (Bauer and Black, 1994; Gerzabek, Kirchmann and Pichlmayer, 1995). It facilitates soil aggregation and provides structural stability - improving air and water relationship - and protects soils from wind and water erosion (Godwin, 1990; Hunt, *et al.*, 1996). It is the source of plant nutrients and Carbon source for micro-flora. Its loss results in reduced infiltration rates, increased crusting, decreased water holding capacity, increased resistance to root penetration, decreased nutrient availability and subsequent degradation of soil structure (Godwin, 1990). Small changes in soil organic matter of soils with low soil organic matter contents - as is the case with the soils under study - are highly significant to the environmental and agricultural potential of these soils (Hunt *et al.*, 1996).

An ideal tillage system should also promote soil water storage, reduce erosion, increase crop yield and be straight forward enough to be adopted by farmers (Cassel, Raczkowski and Denton, 1995). Tillage intensity should be reduced and mulching promoted so that erosion susceptible soils are not exposed to weather conditions (Sauerbeck, 1994). Research has shown that the most cost effective erosion control practices involve keeping crop residues on the surface and reducing tillage as much as possible (Reicosky*et al.*, 1996).

The consequence of inappropriate land-use management is accelerated soil erosion leading to soil degradation and eventually to decreased soil productivity. On-site loss of potential crop production due to eroding away of productive organic-enriched topsoil has always been considered a major threat to sustained food production (Lowery and Larson, 1995). On arable land, the process of sheet erosion is insidious and is usually irreversible. Sheet erosion depletes soil productivity through alteration of soil physical and chemical properties. The extent to which these changes take place greatly depends on the soil type, crop and eco-region (Kaihura*et al.*, 1998).

Sheet erosion is a selective process that deprives the soil of its fine particles, i.e. particle size separation often takes place when soil material is eroded by water. Sediments generally contain a larger amount of the lighter elements, such as humus and higher proportions of finer soil particles than the original soil (Aylen, 1939; Massey and Jackson, 1952; Cormack, 1953; Hudson and Jackson, 1962; Shaxson, 1975; Hanotiaux, 1980; Young, 1980; Elwell and Stocking, 1984 ; Biot, 1986; Elwell, 1987). The finest particles are easily splashed out and/or carried in suspension, while the heavier particles are left behind (Poesen and Savat 1980). The soils are thus impoverished as these nutrient reservoirs are lost together with inherent and applied plant nutrients. The bulk density of the soils is increased and plant available water is decreased. The degree with which particle size separation takes place is higher on sandy soils than on clay soils (Hudson, 1958; 1959).

The major significance of soil erosion therefore, lies in the movement of plant nutrients both inherent and applied (Shaxson, 1975). As a result, the eroded material is enriched with nutrients, organic matter and clay particles. The enrichment ratio, defined as the concentration level of each factor (nutrient element, organic matter, clay) in eroded soil material compared to its level in the soil before erosion (Kejela, 1991), is an important parameter for the assessment of nutrient loss through erosion as well as assessing the impact of erosion on crop productivity. To this end therefore, this chapter seeks to assess the selective process of soil erosion and quantify the nutrient losses with each sediment fraction and the significance of each sediment fraction in carrying plant nutrients during an erosion process.

2. Materials and methods

2.1. Study site

Zimbabwe's climate is moderated by altitude and although the country lies within the tropics its climate is sub-tropical. According to the Koeppen climate classification system, the country is thus classified as temperate Cwb, i.e. mild mid-latitude, with dry winters and hot summers (Roesenberg, 2007). The average temperatures rarely exceed 33^{0}C in summer or drop beyond 7^{0}C in winter (MNTR, 1987). The country has been classified into five agro-ecological regions, namely Natural Regions I, II, III, IV and V. Only Natural Regions I and II have relatively high effective rainfall and are suitable for intensive agricultural production. Natural Regions III, IV and V constitute 83% of the total land area and are not suitable for intensive, high input agriculture (Moyo et al., 1991). Zimbabwe's soils are predominantly derived from granite and the geological complexity of the granites leads to the complexity of the soils (Thompson and Purves, 1978; Nyamapfene, 1991). The clay content of these soils varies according to the degree of weathering (influenced by rainfall) and catenal position (Thompson and Purves, 1978; Nyamapfene, 1991). From among all the soils derived from granite, the sandy soils, of the fersiallitic group, comprise the majority (Thompson and Purves, 1978) and are dominant in the small-holder farming areas (Vogel, 1993). These soils are generally light to medium textured and characterized by the presence of significant amounts of coarse sands (MNRT, 1987; Nyamapfene, 1991). The agricultural potential of these soils is fair (Grant, 1981; MNRT, 1987) and their productivity is likely to decline under intensive continuous cropping (Thompson and Purves, 1978). Therefore increased production can only be achieved through good management as well as application of fertilizers or animal manure (MNRT, 1987).

The research work was carried out at Makoholi Research Station, situated 30 km North of Masvingo town and is the regional agricultural research centre for the sandveld soils in the medium to low rainfall areas. The station lies within Natural Region IV at an altitude of about 1200 m (Thompson, 1967; Anon, 1969). Characteristic of this region is the erratic and unreliable rainfall both between and within seasons (Anon, 1969). Average annual rainfall is between 450 and 650 mm (Thompson and Purves, 1981). The soils at Makoholi are also inherently infertile, pale, coarse-grained, granite-derived sands, (Makoholi 5G) of the fersiallitic group, Ferralic Arenosols (Thompson, 1967; Thompson and Purves, 1978). Arable topsoil averages between 82 and 93% sand, 1 and 12% silt and 4 and 6% clay (Thompson and Purves, 1981; Vogel, 1993). The small amount of clay present is in a highly dispersed form and contains a mixture of 2:1 lattice minerals and kaolinite (Thompson, 1967). The organic matter content is also very low, about 0.8%, while pH ($CaCl_2$) is as low as 4.5. The soils are generally well drained with no distinct structure (Thompson and Purves, 1981), but some sites have a stone line between 50 and 80 cm depth. The low infiltration rates and water holding capacities are due to the soil texture characteristics.

2.2. Experimental design and tillage treatments

The treatments were laid out in a randomized block design replicated three times. The blocks were located at different positions along the slope (Down-slope, Middle-slope and Up-slope). Four different tillage systems were considered namely: conventional tillage, mulch ripping, tied ridging and a bare fallow.

2.2.1. Conventional tillage

The land was ox-ploughed to 23 cm depth, soon after harvest (winter ploughing), using a single-furrow mould-board plough and thereafter harrowed with a spike harrow in spring. All crop residues were removed from the plots, as is the practice in the communal areas. This tillage system is the most commonly used tillage system in the communal areas and was chosen as a standard primary tillage method, i.e. including this treatment provides a baseline for assessing the merits of other treatments (Working Document, 1990).

2.2.2. Mulch ripping

Crop residues from the previous season were left to cover the ground and only rip lines, 23 cm deep, were opened between the mulch rows, using a ripper tine. The rip lines acted as crop rows and were alternated every year, to allow roots ample time to decay. Two basic conservation tillage components were used here, i.e. minimum tillage and mulching. The main aim was to maximize infiltration through rainfall interception provided by the mulch, thus minimizing run-off. According to Hudson (1992), this parameter is the most important in the semi-arid regions, where soil moisture is the most limiting factor in agricultural production. This treatment is one of the basic conservation tillage systems, which has shown great potential in protecting the soils, without compromising the production potential and is currently being promoted by the Institute of Agricultural Engineering.

2.2.3. Tied ridging

The land was ploughed to the recommended depth of 23 cm in the first year and crop ridges constructed at 1 in 250 grade, using a ridger. The ridges were about 900 mm apart and small ties were put at about 700-1000 mm along the furrows between the crop ridges. These ties were between one half to two thirds the height of the crop ridges allowing for the water to flow over the ties and not over the ridges (Elwell and Norton, 1988). The ridges were maintained several years through re-ridging so as to maintain their correct size and shape. This treatment has been found to reduce run-off, and the soil losses are also reduced to satisfactorily low levels of 0.1 to 0.3 t/ha, much less than the tolerable limit of 5 t/ha/yr. (Elwell and Norton, 1988).

2.2.4. Bare fallow

Ploughing, up to 23 cm depth, was done using a tractor disc plough and disc harrow. The plots were kept bare and weed free, by spraying the germinating weeds during the season. This treatment is important for soil erodibility assessment and modeling purposes, as it gives the highest possible soil loss values and will probably give the lowest nutrient loss values as no fertilizers are applied.

At the beginning of this study, all trial plots had been under cultivation and the same treatment for a period of five years, having been opened up from virgin woodland. All tillage operations were carried out soon after harvest before the soil dried out. Shortly before the on-set of the rains, planting holes were made on all crop treatments, using a hand-hoe. Thereafter basal fertilizer and a nematicide were applied into the planting holes.

2.3. Agronomic details

Maize (*Zea mays* L.) is the staple food in Zimbabwe. For this reason, maize was chosen as a trial crop, so as to make the research project relevant to the small holder areas. Due to the dry conditions prevailing at Makoholi, maize variety R 201, which tolerates moisture stress and is short seasoned, was used. The crop spacing of 900 mm inter-row and 310 mm in-row were used resulting in a plant population of about 36 000 plants/ha. All weeding operations were done using a hand-hoe. The problems of nematodes, very common in the sandy soils and that of maize stalk borer were controlled, so as to minimize the influence of factors other than those imposed by treatments. Carbofuran, a nematicide was applied into the planting holes before the on-set of the rains, while Thiodan (against maize stalk borer) was applied six weeks after planting.

On all plots planting holes of about 10 cm depth and diameter were opened before the onset of the rains. Thereafter Carbofuran, was applied into these planting holes at a rate of 20 kg/ha. Compound D (N:P:K = 8:14:7) was also applied into the planting holes at a rate of 200 kg/ha to give a final ratio of 16 kg N: 12 kg P: 12 kg K. The nematicide and fertilizer were then slightly covered with soil and left until adequate rainfall had been received.

Once the profile of the ridges was wet throughout, maize was planted, two seeds per station. Ten days after planting, crop emergence count was carried out followed by weeding. The crop was then thinned out to one plant per station. When the crop was about six weeks, ammonium nitrate top-dressing fertilizer was applied at 100 kg/ha, amounting to 34.5 kg N/ha. The ammonium nitrate application coincided with the second weeding and the application of Thiodan, to control maize stalk borer.

2.4. Soil loss assessment

The standard soil erosion methodology for Zimbabwe (Wendelaar and Purkis, 1979) was used, where the plots were laid out at 4.5% slope. Soil loss and run-off measurements were from 30 m x 10 m run-off plots, with 5 m border strips on either side. The length of the plots was orientated up-slope. Tillage operations were done across the slope. Polythene strips were dug in to form the boundary around each 300 m^2 plot (Working Document, 1990). For the tied ridging treatment, the collection area was 150 m long and 5 crop rows wide (4.5 m), with 2 guard rows above and below. The crop ridges were laid at 1% slope and the length of the plots was orientated across the slope. Surface run-off and soil loss from each plot were allowed to collect in a gutter at the bottom of the plot. From the gutter these were channeled through a PVC delivery pipe into the first 1500 litre conical tank. The collection tanks were calibrated and run-off was measured using a metre-stick. Once the first tank was full its overflow passed through a divisor box with ten slots, which channeled only one tenth of the overflow into the second tank. Nine tenths of this overflow was allowed to drain away, thus increasing the capacity of the second tank. Due to the larger net plots of the tied ridging treatment, three tanks were installed, so as to capture the anticipated larger volume of sediments.

2.5. Sampling eroded material

Tanks were emptied at the end of each storm unless the interval between storms was too short to allow emptying. Sediments and run-off (including the suspended material) collected from run-off plots were treated as two different entities. Suspension was pumped out and sub-sampled for the determination of soil concentration in run-off, using the Hach spectrophotometer DL/2000. Later the sludge was transferred into 50 l milk churns, topped up with water to a volume of 50 litres and weighed. The mass of oven dry soil, M_o (kg) was calculated using the following equation (Wendelaar and Purkis, 1979; Vogel, 1993):

$$M_o = 1.7x(M_s - M_w) \tag{1}$$

Where M_s = mass of fixed volume of sludge (kg)

M_w = mass of the same volume full of water (kg)

1.7 = constant for the soil type

For clay, organic matter and plant nutrient assessment of the eroded soil, the collected sediments were thoroughly mixed and a sample taken by driving a hollow plastic tube into the sludge "profile" in the churn. Suspension was pumped out into 55 litre plastic containers, left to stand for 3 days, a water sample taken and the settled material at the bottom of the container sampled. Both soil samples were then air dried and analyzed individually i.e. for each storm, thus the averages given for the different treatments refer to twenty-one effective storms recorded during the season 1; nine storms during season 2 and twenty-two storms in season 3.

2.6. Soil sampling

Soil sampling on trial plots was carried out at the end of each season. Composite soil samples were taken (8-10 samples per plot) within the plough depth of 0-250 mm, using a split auger. They were then air dried and sieved.

2.7. Laboratory analysis

An analysis of the sediments for macro-nutrients was carried out, where the different sediment fractions (water, suspended material and sludge) were treated as different entities. The main aim being to quantify nutrient losses as a result of erosion and to ascertain which sediment component carries the most nutrients. Total nutrients were determined in an effort to capture all forms of nutrients and therefore give a clear picture of how much was lost with erosion, rather than giving a mere fraction of the available form. Soil samples from the trial plots, as well as eroded material, were analyzed using the following methods:

2.7.1. Texture

Texture was determined using the hydrometer method as described by Gee and Bauder (1986), where 100g of air dried soil in 15 ml of calgon and 500 ml of water were stirred for 15 minutes using an electrical stirrer. The mixture was then transferred into 1 litre cylinders and diluted with water to 1 litre. After shaking the cylinder, time and temperature readings were taken and hydrometer readings were taken after 5 minutes (clay and silt) and five hours (clay). The sand fraction was determined by transferring the contents of the cylinder on a 50 micron sieve and washing away all the silt and clay fractions and then drying.

2.7.2. Organic carbon

The Walkley and Black method as described by Nelson and Sommers (1982) was used. One g of soil was digested with 10 ml of 1N potassium dichromate solution and 20 ml of concentrated sulphuric acid. After ten minutes 100 ml of water were added, the mixture shaken and then read on a spectrophotometer.

2.7.3. Total nitrogen

Nitrogen was determined using the microkjeldahl method as described by Bremner and Mulvaney (1982). The methodology, in brief, was as follows: The soil was digested with concentrated sulfuric acid and hydrogen peroxide in the presence of a selenium catalyst. Organic nitrogen was converted into ammonium sulfate. The solution was made alkaline and the liberated ammonia (NH_3) was distilled and trapped in boric acid. The boric acid was titrated with a standard mineral acid.

2.7.4. Total phosphorous

The ignition method as described by Olsen and Sommers (1982)was used. Air dried soil was weighed into a crucible and the crucible placed into a muffle oven. The sample was ignited at 500 - 600 ^{0}C for three hours after which it was allowed to cool. Sulfuric acid was added and the mixture shaken on a reciprocating shaker for three hours. The mixture was filtered and 0.5 ml of 3M sulphuric acid were added to 5 ml of the aliquot. Twenty ml of water were added together with 4 ml of Reagent P and ascorbic acid. After 20 minutes P absorbance was measured.

2.7.5. Total potassium

The wet digestion method using perchloric acid as described by Knusden, Peterson and Pratt, (1982) was used. The mixture of finely ground soil, hydrofluoric acid and perchloric acid was heated and cooled. Some more hydrofluoric acid was added and the contents were evaporated in a sand bath. After cooling, 6N HCl and water were added and the mixture further heated until it boiled gently. The contents were transferred to a flask, diluted to volume and filtered. K was read from a flamephotometer.

2.7.6. *Nutrients dissolved in run-off*

Run-off was filtered and the aliquot treated as soil extract, where the nutrient concentration was either titrated with boric acid, for N determination, read from an Atomic Absorption Spectrophotometer for the determination of P or read from a flamephotometer in the case of K.

2.8. Statistical analyses

The differences in soil loss, run-off, plant growth parameters and yield attributed to treatment were analyzed with the analysis of variance (ANOVA) procedure of Genstat 5 Release 1.3 statistical package. An independent t-test was used to compare the means of different populations. Unless otherwise indicated, significance is indicated at $P < 0.05$ (*), 0.01(**) to 0.001 (***).

3. Results

3.1. Run-off

As slope steepness and slope length are the same for all treatments, run-off is thus expected to be mainly dependent on the amount, distribution and intensity of seasonal rainfall, infiltration rate, which is directly influenced by tillage and ground cover. A tillage system that either maintains a good soil structure, or inhibits run-off velocity and raindrop impact, or forces water to pond, or has a good ground cover (mulch or crop) tends to have a higher infiltration rate and therefore, lower run-off.

The highest run-off was recorded under bare fallow, with a run-off range of 17 – 39% of total seasonal rainfall. Conventional tillage recorded the second highest average run-off ranging from 13 to 22% of total seasonal rainfall. This could be attributed to a somewhat better infiltration rate at the beginning of the season, as the soil would be loose. The best treatments in conserving water were mulch ripping and tied ridging, which had run-off ranges of 9 - 15% and 1 - 11% respectively. The mulch cover has all the positive attributes that have been highlighted, i.e., reducing raindrop impact thus inhibiting soil capping, reducing run-off velocity and increasing water infiltration. Under tied ridging, run-off is also contained at low levels by way of water ponding. The micro-dams force water to pond - thus increasing infiltration - until all the micro-dams are full and start overtopping along the ridges, allowing very little water to leave the system.

Table 1 shows ANOVA results between treatments and as influenced by year. Note that the year x treatment interaction was mainly due to the differences in rainfall amount. Run-off differed significantly between treatments at $P < 0.001$. To properly evaluate the effectiveness of the conservation tillage treatments, the mean of conventional tillage versus the mean of the two conservation tillage treatments was compared using an independent t-test. The results of this test are given in Table 1. Despite the overall high significant variation between the treatments, it was established that this difference was only between conventional tillage and the two conservation tillage treatments (mulch ripping and tied ridging). There was, however no significant difference between the two conservation tillage treatments. This finding confirms that both mulch ripping and tied ridging treatments are effective in reducing run-off when compared to conventional tillage.

Treat/Year (Rainfall)	Year 1 (483 mm)	Year 2 (384 mm)	Year 3 (765 mm)	Overall mean (mm)	Source of variation	Run-off
CT	94.9	48.7	169.5	104.4	Treat	***
MR	4.6	3.6	111.4	39.8	Year	***
TR	16.0	4.5	81.8	34.1	Treat x Year	NS
BF	122.7	65.3	295.3	161.1	MR vs TR	NS
Overall mean	59.5	30.5	164.5	84.9	CT vs (MR, TR)	***
n = 9 (Treatment)	s.e.d. = 8.07	$s^2 = 340.4$			Yr 1 vs Yr 2	*
n = 12 (Year)	s.e.d. = 7.53	df = 24			Yr 3 vs (Yrs 1, 2)	***
n = 3 (Treatment x Year)	s.e.d. = 15.06					

Key: CT = Conventional Tillage; MR = Mulch Ripping; TR = Tied Ridging; BF = Bare Fallow

Table 1. Run-off (mm) as affected by tillage and year (rainfall) and their interactions at MakoholiContill site during three seasons

The amount of run-off recorded during the different years also differed significantly at $P < 0.001$. This was due to the high variation in rainfall amounts received during the three seasons. Year 1 received close to twice the rainfall amount received during Year 2. Run-off increased by more than six times, due to the concentration of rainfall in January, inducing saturated conditions, which led to high run-off. As a result of this highly significant seasonal variation an independent t test was carried out on the means of the different years. The results showed that the 100 mm difference between Year 1 and Year 2 resulted in significantly different run-off levels, at $P < 0.05$, while run-off from Year 3 differed significantly ($P < 0.001$) from the mean of that of Year 1 and Year 2. There was no significant difference for the interaction between the treatment and the year ($P = 0.145$).

The significant difference between the years further prompted an analysis of variance to establish how treatments varied within the individual years (Table 1). The overall run-off treatment differences were significant at $P < 0.01$ for the Year 1 and Year 3. A higher overall significant treatment difference was found for Year 2 indicating that the differences in run-off become more pronounced if seasonal rainfall amount was low than during wetter seasons. During wet seasons, run-off was also higher under the conservation tillage systems as they reached saturation point faster due to the already high residual soil moisture. An independent t-test showed that conventional tillage differed highly significantly from the mean of conservation tillage treatments throughout the three seasons. Mulch ripping and tied ridging, however did not differ significantly in any one of the seasons. This finding further emphasizes the water conservation potential of mulch ripping and tied ridging and also shows that a lot more rain water is lost under conventional tillage.

3.2. Soil loss

Soil losses followed the same trend as rainfall, especially under the bare fallow, where there was no ground cover (Table 2). The highest soil losses were recorded under bare fallow, averaging 93 t/ha/yr. Soil losses under conventional tillage averaged 34 t/ha/yr, while mulch ripping and tied ridging recorded soil loss averages of 1.7 and 3.3 t/ha/yr respectively. The importance of crop cover on soil erosion is shown by the different cropped treatments, especially conventional tillage, where the reduction in erosion (34 t/ha/yr from that of bare fallow, 93 t/ha/yr) is attributed to cover alone and not tillage system. Overall, the treatments differed significantly at $P < 0.001$. Independent t-tests showed that conventional tillage differed highly significantly from the two conservation tillage treatments, while there was no significant difference between the conservation treatments. This finding, tallies with the run-off results and is in accordance with expectations as soil loss is a function of run-off.

Treat/Year (Rainfall)	Year 1 (483mm)	Year 2 (384mm)	Year 3 (765mm)	Overall mean (t/ha)	Source of variation	Soil loss
CT	40.2	6.8	54.0	33.7	Treat	***
MR	0.2	0.1	4.8	1.7	Year	***
TR	3.0	0.1	3.5	2.2	Treat x Year	***
BF	84.1	43.5	152.5	93.4	MR vs TR	NS
Overall mean	31.9	12.6	53.7	32.7	CT vs (MR, TR)	***
n = 9 (Treatment)	s.e.d. = 4.00	$s^2 = 71.83$			Yr 1 vs Yr 2	***
n = 12 (Year)	s.e.d. = 3.46	df = 24			Yr 3 vs (Yrs 1,2)	***
n = 3 (Treatment x Year)	s.e.d. = 6.92					

Key: CT = Conventional Tillage; MR = Mulch Ripping; TR = Tied Ridging; BF = Bare Fallow

Table 2. Soil losses (t/ha) as affected by tillage and year (rainfall) and their interactions at MakoholiContill site during three seasons

Year had a significant effect on soil loss ($P < 0.001$) due to the varied seasonal rainfall totals. The Years 1 and 2 varied significantly at $P < 0.001$. When the mean of these two seasons was compared with the mean of Year 3, the difference also varied significantly at $P < 0.001$. The influence of rainfall on soil loss is apparent, as the season with the highest rainfall also recorded the highest soil loss and vice versa. The analysis of variance during the individual years, gave a significant overall treatment difference at $P < 0.001$ across all years. A comparison between the treatment means also confirmed a significant variation ($P < 0.001$), between conventional tillage and the mean of mulch ripping and tied ridging. There was no significant difference between mulch ripping and tied ridging. As soil loss is a function of run-off, the increase of soil loss with the increase in the number of years of cultivation was expected and the same range of factors that affected run-off should be responsible for these increases in soil loss.

3.3. Particle size distribution of the sediments

The average mechanical composition of the sediments collected over the three years is shown in Table 3. Only clay and silt fractions are given. The sediments from the conservation tillage treatments comprised of more clay, i.e. more suspended material as compared to sludge (coarse material), while under the conventional tillage systems more sludge was lost compared to suspended material. There was very little clay/ silt found in sludge, while the suspended material hardly contained any sand fraction, i.e. over 90% of the suspended material was found to be clay and silt fractions. It is clear that the suspended material comprises of the most reactive soil particles (clay, silt and organic matter) and thus its loss is most detrimental to the soils' productivity as compared to sludge. Furthermore, the total sediments (sludge + suspended material) had higher clay and silt contents when compared to the original soil.

The ratios between these two sediment components were worked out for the different tillage systems (Table 3). The results show that 10 - 17 times more sludge than suspended material was found under the bare fallow, while the ratio ranged between 1.5 and 5 under conventional tillage and below 1, under the conservation tillage treatments (mulch ripping and tied ridging). This is an indication that not so much soil is moved during erosion under these treatments, while under bare fallow, mass movement is realized. The impediments created under the two conservation tillage systems ensured that the run-off velocity was reduced thus allowing no sheet wash but only the suspended soil particles to leave the system.

Figure 2 shows the actual amount of clay lost with sludge, suspension and with sediments as a whole. Although there was a lower percentage of suspended material as compared to sludge, under the bare fallow and conventional tillage, the actual amount of clay lost with suspension exceeds that lost with sludge. The clay amount lost with sediments showed the following trends:

- the highest amount of clay was lost under the bare fallow (an average of 7 t/ha/yr.) followed by conventional tillage (5 t/ha/yr.) and only a negligible amount was lost under two conservation tillage treatments, i.e. 0.9 and 0.8 t/ha/yr. for mulch ripping and tied ridging respectively

- the amount of clay lost with suspension followed the same trend as that of the total sediments although the differences between bare fallow and conventional tillage were relatively smaller (Figure 1). The significant difference among the treatments was realized in the clay amount lost with sludge.

Year/Treat.	Soil loss (t/ha)		Ratio	Clay content (%)		Silt content (%)	
	Sludge	Susp	Slud:Susp	Sludge	Susp	Sludge	Susp
Year 1 (483 mm)							
BF	74.54	7.28	10.24	1.32	58.00	2.93	37.87
CT	27.97	6.33	4.42	1.22	57.69	3.21	34.40
MR	0.00	0.17	0.00	0.00	59.93	4.10	36.75
TR	0.38	1.16	0.33	0.81	66.68	4.59	33.32
Year 2 (384 mm)							
BF	41.09	2.37	17.34	2.78	52.56	1.58	29.55
CT	4.12	2.70	1.53	2.79	57.81	2.69	36.45
MR	0.00	0.08	0.00	0.00	59.75	0.00	38.02
TR	0.04	0.10	0.40	2.01	69.95	1.76	43.57
Year 3 (765 mm)							
BF	139.02	13.36	10.41	3.02	64.89	1.48	27.66
CT	45.66	8.44	5.41	3.15	76.97	2.22	34.27
MR	1.84	3.03	0.61	4.00	83.43	2.44	17.25
TR	0.65	2.88	0.23	4.02	83.40	4.12	6.05

Key: CT = Conventional Tillage; MR = Mulch Ripping; TR = Tied Ridging; BF = Bare Fallow

Table 3. Relationship between sludge and suspended material in erosion sediments from four tillage systems over three years at MakoholiContill site

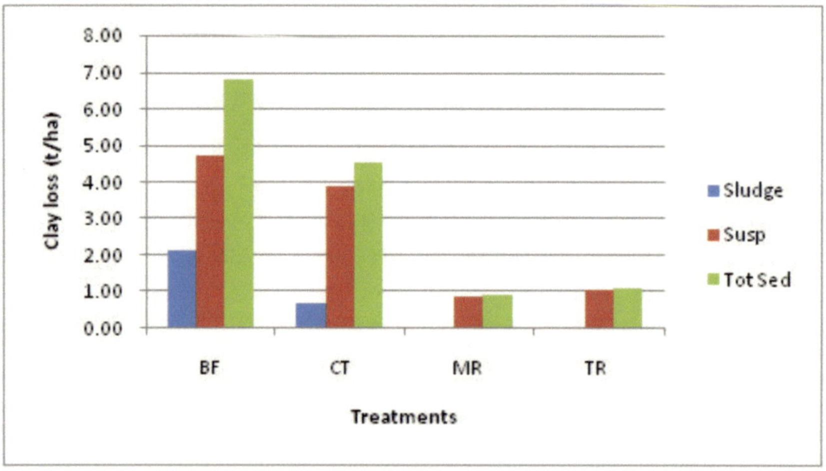

Key: CT = Conventional Tillage; MR = Mulch Ripping; TR = Tied Ridging; BF = Bare Fallow

Figure 2. Average clay loss with sediments during three years at MakoholiContill site with sludge, suspension and total sediments

The clay enrichment ratios for the total sediments (clay content in soil: clay content in sediments) show that the sediments have distinctly more clay than the original soil (Table 4). In all cases the bare fallow had enrichment ratios of less than 2, while under conventional tillage the ratio ranged between 2.8 and 6.0. The two conservation tillage treatments recorded the highest enrichment ratios, as expected, of between 12.4 and 14.5 for mulch ripping and 13.8 - 19.0 for tied ridging. The very high clay enrichment ratios found under conservation tillage treatments indicate the very low run-off velocity which only carries suspended material but has not enough energy to erode and carry coarse particles, as is the case under conventional tillage and the bare fallow. Table 4 also shows that clay content in the sediments varied highly significantly between the treatments ($P < 0.001$), while there was no difference ($P = 0.966$) between the sediment composition over the years

Treat./Year	**Clay in total sediments %**	**Clay t/ha**	**Clay Enrichment Ratio**
Year 1			
BF	6.36	5.20	1.41
CT	11.63	3.99	2.78
MR	58.83	0.10	13.68
TR	50.00	0.77	13.81
Year 2			
BF	5.50	2.39	1.22
CT	25.07	1.67	5.98
MR	62.50	0.05	14.53
TR	50.00	0.07	13.81
Year 3			
BF	8.45	12.87	1.87
CT	14.68	7.94	3.50
MR	53.39	2.60	12.42
TR	68.84	1.43	19.02
Source of variation	Clay % in sediments	Clay (t/ha)	
Treatment	***	***	
s.e.d.	5.65	2.676	
Year	NS	***	
s.e.d.	19.23	2.320	

Key: CT = Conventional Tillage; MR = Mulch Ripping; TR = Tied Ridging; BF = Bare Fallow

Table 4. Clay loss with sediments and its enrichment ratios for the different tillage systems during three seasons at MakoholiContill site

3.4. Organic matter loss with sediments

The original organic matter content of the virgin soils averaged approximately 0.8%, (Table 5). After continuous cultivation for five years the organic matter content was found to have declined by 25% under the bare fallow; 19% under the conventional tillage; 6% under mulch ripping and 9% under tied ridging. This finding shows that with continuous cultivation the organic matter status of these soils decreases, more so if no plant residues are left in the field, e.g. bare fallow. The higher organic matter content under conventional tillage, compared to bare fallow, is a result of roots left behind after harvest. Tied ridging combines this effect with that of soil conservation to give an even better maintenance of organic matter. The best effect is, however, achieved under mulch ripping, where roots together with plant residues and soil conservation effects contribute to better organic matter maintenance, thus only 6% was lost. The mineralization of organic matter after cultivation is expected to take place but by further addition of mulch the depreciation rate is lowered drastically. Reduced tillage in the mulch ripping treatment, as compared to other treatments, further contributes to conservation of organic matter.

Treatment	Virgin land OM %	After five of cultivation OM %	OM reduction %
BF	0.72	0.53	24.9
CT	0.84	0.68	18.9
MR	0.85	0.80	5.7
TR	0.70	0.64	9.1

Key: CT = Conventional Tillage; MR = Mulch Ripping; TR = Tied Ridging; BF = Bare Fallow

Table 5. Soil organic matter content of the soils (0 - 25 cm depth) for different tillage systems as at opening from virgin land and five years later at MakoholiContill site

The concentration of organic matter in sediments was higher for conservation tillage systems than for conventional tillage and bare fallow. This resulted in higher enrichment ratios (organic matter content in soil: organic matter content in sediments) for conservation tillage systems. Mulch ripping and tied ridging recorded enrichment ratios of 4.6 and 3.4 respectively, while conventional tillage recorded an enrichment ratio of 2.8 and bare fallow 2.6 (Table 6). The total amount of organic matter lost with conservation tillage was, however, only a fraction of that lost from conventional tillage and bare fallow. Under bare fallow an annual average of 424 kg/ha was lost, while under conventional tillage 299 kg/ha were lost, mulch ripping lost only 55 kg/ha and tied ridging 61 kg/ha/yr (Table 7). It is quite obvious that although the organic matter concentrations in sediments of bare fallow and conventional tillage are quite low, the extensive losses of soil contributed to a tremendous total loss. Higher contents for conventional tillage as compared to bare fallow show the contribution of roots to the soil organic matter, thus a higher depreciation is apparent under bare fallow, where no crops are grown.

Under the conventional tillage treatments organic matter losses were much higher than with conservation tillage. Whereas most organic matter in conservation tillage was lost in suspension, losses with conventional tillage were more evenly distributed between suspension and sludge, because of the very high sludge losses, see Table 7.

Treatm.	**OM content (%)**		**Enrichment ratio**
	Soil	**Sediments**	
Year 1			
BF	0.54	1.63	3.02
CT	0.68	2.24	3.29
MR	0.80	5.36	6.70
TR	0.63	2.82	4.48
Year 2			
BF	0.54	1.05	1.94
CT	0.68	1.27	1.87
MR	0.80	2.88	3.60
TR	0.63	1.22	1.94
Year 3			
BF	0.54	1.49	2.76
CT	0.68	2.13	3.13
MR	0.80	2.69	3.36
TR	0.63	2.34	3.71

Key: CT = Conventional Tillage; MR = Mulch Ripping; TR = Tied Ridging; BF = Bare Fallow

Table 6. Organic matter contents of the soils and sediments and calculated enrichment ratios for four tillage treatments, over three seasons at MakoholiContill site

The amount of organic matter lost varied significantly ($P < 0.001$) among all treatments. Contrasting the different systems against one another showed that conventional tillage did not differ significantly from the bare fallow. The mean of conventional tillage and bare fallow, however, differed significantly at $P < 0.001$, with that of mulch ripping and tied ridging, indicating that the two conservation tillage treatments are very effective in conserving and/or maintaining soil organic matter. When conventional tillage was also compared to the mean of the conservation tillage treatments, the difference was significant, $P < 0.001$. Finally the two conservation treatments were compared within the group and they were not significantly different. As it is important to show which of the two conservation treatments performs better than the other, an independent t-test was carried out, i.e., disregarding the other two treatments altogether and comparing the two conservation treatments only. The results showed a significant difference at $P < 0.05$, where lower losses were found under the mulch ripping treatment.

Treatm.	OM content %		OM loss kg/ha		Enrichment ratio	
	Sludge	Susp.	Sludge	Susp.	Sludge	Susp.
Year 1						
BF	0.23	2.59	168.25	188.86	0.43	4.80
CT	0.35	3.27	99.15	206.78	0.51	4.81
MR	0.00	5.02	0.00	8.46	0.00	6.28
TR	0.42	4.56	1.58	52.74	0.67	7.24
Year 2						
BF	0.04	2.06	11.04	48.20	0.07	3.81
CT	0.10	2.44	4.60	70.15	0.15	3.59
MR	0.00	2.88	0.00	2.58	0.00	3.60
TR	0.15	2.29	0.06	2.11	0.24	3.63
Year 3						
BF	0.37	2.61	509.49	345.07	0.69	4.83
CT	0.45	3.81	208.55	307.09	0.66	5.60
MR	0.31	5.07	4.80	148.53	0.39	6.34
TR	0.55	4.12	4.24	121.30	0.87	6.54
ANOVA						
Treatment	***	***	***	***		
Year	***	***	***	***		

Key: CT = Conventional Tillage; MR = Mulch Ripping; TR = Tied Ridging; BF = Bare Fallow

Table 7. Differences in the organic matter contents of sludge and suspended soil over three seasons at MakoholiContill site

Soil organic matter is generally associated with the finer and more reactive clay and silt fractions of the soil (Follet*et al.*, 1987). It is, therefore as expected that more organic matter should be lost with suspended load than with sludge. Table 7 shows the two sediment parameters (sludge and suspended load), which were treated as different entities. In relation to this, organic matter contents, quantities and enrichment ratios for the different parameters are given.

Under the conventional tillage treatments relatively less soil was lost as suspended load. The amount of organic matter lost with this fraction was, however, substantial, i.e. 40 - 81%

http://dx.doi.org/10.5772/3402

(BF) and 60 - 94% (CT) of the total organic matter lost. The soil lost under mulch ripping was almost entirely in suspended form. This resulted in almost all the organic matter (97 - 100%) being lost with suspended material. Under tied ridging most of the soil was also lost as suspension, 75% of the total soil loss, with 97% of total organic matter loss. The enrichment ratios are thus very high for the suspended soil consisting mainly of clay and silt. The sludge has lower enrichment ratios and the organic matter contents are even less than of the original soil.

Many scientists have reported on the selective nature of soil erosion (Aylen, 1939; Hudson and Jackson, 1962; Shaxson, 1975; Elwell, 1987; Lal, 1988 among others), however the extent to which the soils have been impoverished or the sediments enriched have mainly been estimated. This study showed that sheet erosion is selective as sediments recorded higher contents of clay, silt and organic matter than the original soil. The particle size distribution of sludge was mainly coarse sand and had a maximum of 4% clay. Suspended material on the other hand had up to 83% clay, the rest mainly being the silt fraction. Sediments from conservation tillage systems comprised of more sand than the clay fraction, bare fallow had 13 times and conventional tillage 4 times less sludge than suspension. Due to the high losses of coarse material under the conventional tillage, the clay enrichment ratio of sediments was lower than under the conventional tillage systems. The bare fallow recorded only 1.5 and conventional tillage 4.1 times more clay in the sediments. Under the conservation tillage systems more soil was lost in suspension than as sludge. Mulch ripping recorded 0.2 times and tied ridging 0.3 times more sludge than suspension, resulting in higher clay enrichment ratios of 13.5 and 15.5 under mulch ripping and tied ridging respectively. However, these high enrichment ratios amounted to less amount of clay lost from conservation tillage compared to conventional tillage, due to the reduced total sediments lost under the conservation tillage systems. Under bare fallow an average of 7 t/ha of clay were lost, 5 t/ha under conventional tillage and 0.9 and 0.8 t/ha under mulch ripping and tied ridging respectively. The very high enrichment ratios under the conservation tillage systems are a reflection of the soil losses, which were lost mainly as suspended material, which constitutes fine soil particles, however the very low amount of soil led to negligible losses of total clay loss with sediments.

This selective nature of erosion manifested itself in the high enrichment ratios of the sediments as compared to the original soil. This means that soil fertility is affected severely by the soil lost in suspension, as it constitutes mainly of clay and organic matter fractions, which are the main sources of nutrients (Stocking, 1983). The soil structure and water holding capacity are also affected as these soil fractions are responsible for soil aggregation and influence the water dynamics of the soil (Follet*et al.*, 1987; Stocking and Peake, 1987). The sludge fraction, however, affects mainly the soil productivity through reduction in soil tilth as it contains few reactive particles.

As was established for the clay loss, the organic matter enrichment ratios were higher under mulch ripping (4.6) and tied ridging (3.4) as compared to conventional tillage (2.8) and bare fallow (2.6). Exceptionally high organic matter losses were realized under conventional tillage systems as compared to conservation tillage systems, due to the high sediment losses.

The bare fallow lost an annual average of 424 kg/ha, while conventional tillage lost 299 kg/ha, mulch ripping lost only 55 kg/ha and tied ridging 61 kg/ha/yr and the treatments differed significantly from one another. The proximity and concentration of soil organic matter near the soil surface (< 250 mm) and its close association with plant nutrients in the soil makes erosion of soil organic matter a strong indicator of overall plant nutrient losses resulting from erosion (Follet*et al.*, 1987). Thus the effectiveness of the two conservation tillage treatments can be appreciated based upon the small amount of organic matter lost with eroded sediments, compared to the conventional tillage.

In situ measurement of organic matter as a measure of soil erosion yielded fruitful as the organic matter levels dropped drastically after five years of cultivation, especially under the conventional tillage systems. The bare fallow lost 25%, conventional tillage 19%, tied ridging 9% and mulch ripping 6% of total organic matter found on virgin land. This is in agreement with the very high losses of fine particles lost under the conventional tillage systems as compared to conservation tillage. It is apparent that through conservation of the soil under mulch ripping and tied ridging, the organic matter status in the soil is maintained, thus the soil structure and soil productivity.

As most of the soil fertility is associated with clay and humus and these also affect microbial activity, soil structure, permeability and water holding capacity (Troeh*et al.*, 1980), it is clear that through sheet erosion the land is degraded chemically, physically and biologically. Thus not only soil fertility is reduced, but also soil productivity, which unlike fertility cannot be addressed by mere fertilizer application.

3.5. Nutrient losses with sediments

Before the assessment of the nutrient losses with erosion, it was important to evaluate the nutrient (N, P, K) status of the soils. From Table 8, it is apparent that the most abundant nutrient in the soil is potassium, followed closely by nitrogen and the least abundant is phosphorus. Since total nutrients are considered it is expected that the nutrient with the highest concentration in the soil will also result in the highest losses and vice versa. Thus, comparing the amount of different nutrients lost with the sediments may not be very meaningful but a method of evaluating and comparing the loss of different nutrients should also be based relatively upon the status of that nutrient in the soil. This method involves the determination of nutrient concentration in the soil and in the sediments and calculating the enrichment ratios.

The nutrient losses were calculated using the following equation:

$$Nut_{los} = Soil_{los} \times Nut_{conc} \tag{2}$$

where Nut_{los}= any nutrient lost with sediments (kg/ha)

$Soil_{los}$= mass of soil lost by erosion (kg/ha)

Nut_{conc}= the concentration of a nutrient in the sediment (ppm or %)

Treatment	Nutrient status of the soil		
	N %	P ppm	K ppm
BF	0.04	39.4	554.2
CT	0.05	52.0	616.7
MR	0.05	62.2	575.0
TR	0.05	91.8	487.5

Key: CT = Conventional Tillage; MR = Mulch Ripping; TR = Tied Ridging; BF = Bare Fallow

Table 8. Nutrient of the soils as at beginning of the study at Makoholi Contill site

3.5.1. Nutrient losses with total sediments

Nitrogen

Using Equation 2 to calculate the amount of N lost with erosion, the highest total nitrogen losses were realized under bare fallow, at 28 kg/ha followed by conventional tillage (16 kg/ha), while they were least under mulch ripping (2.3 kg/ha), which was also barely different from tied ridging (2.7 kg/ha), see Table 9. Total nitrogen loss differed significantly ($P < 0.001$) between the different treatments, different years and for the treatment x year interaction. These results follow, as expected, the same trend that was established for soil loss (Table 2) and serve to confirm the dependence of nutrient losses with the amount of soil lost from a field. The maintenance of soil under the two conservation tillage treatments is also directly related to the lower N losses. Although nitrogen losses were highest under the bare fallow, the actual nutrient concentration in the soil was least under this treatment (Table 12) because no fertilizer was applied and the sediments under this treatment comprised mainly the non-reactive coarse particles.

Treat/Year (Rainfall)	Year 1 (483 mm)	Year 2 (384 mm)	Year 3 (765 mm)	Overall mean	Source of variation	N loss
CT	17.40	6.82	23.22	15.81	Treat	***
MR	0.53	0.16	6.06	2.25	Year	***
TR	3.03	0.15	4.92	2.70	Treat x Year	***
BF	32.10	9.06	44.10	28.42	MR vs TR	NS
Overall mean	13.27	4.05	19.58	12.30	CT vs (MR, TR)	***
n = 9 (Treatment)	s.e.d. = 1.341	$s^2 = 8.097$				
n = 12 (Year)	s.e.d. = 1.162	df = 24				
n = 3 (Treatment x Year)	s.e.d. = 2.323					

Key: CT – Conventional Tillage; MR = Mulch Ripping; TR = Tied Ridging; BF = Bare Fallow

Table 9. Total nitrogen loss (kg/ha) as a result of erosion under different tillage systems over three years at MakoholiContill site

Phosphorus

The overall phosphorus loss of 0.5 kg/ha, was as expected, much lower than nitrogen loss (12.3 kg/ha), due to the generally low P status in the sandy soils. The bare fallow had the highest P loss of 0.9 kg/ha followed by conventional tillage with 0.8 kg/ha, tied ridging 0.2 kg/ha and the least P losses were recorded under mulch ripping (0.09 kg/ha) (Table 10). This trend was to be expected, as nutrient losses are a function of soil loss. Despite the low losses, the treatments and years gave highly significant differences at P < 0.001. The two conservation tillage treatments were not significantly different from one another.

Treat/Year (Rainfall)	Year 1 (483 mm)	Year 2 (384 mm)	Year 3 (765 mm)	Overall mean	Source of variation	P loss
CT	1.403	0.182	0.666	0.750	Treat	***
MR	0.057	0.009	0.208	0.091	Year	***
TR	0.282	0.008	0.218	0.169	Treat x Year	***
BF	1.269	0.245	1.069	0.861	MR vs TR	NS
Overall mean	0.753	0.111	0.540	0.468	CT vs (MR, TR)	***
n = 9 (Treatment)	s.e.d. = 0.0667	s^2 = 0.02004 df = 24				
n = 12 (Year)	s.e.d. = 0.0578					
n = 3 (Treatment x Year)	s.e.d. = 0.1156					

Key: CT = Conventional Tillage; MR = Mulch Ripping; TR = Tied Ridging; BF = Bare Fallow

Table 10. Total phosphorus (kg/ha) as a result of erosion under different tillage systems over three years at MakoholiContill site

Potassium

Potassium was, as expected, lost in greater quantities when compared to the other elements (overall 17.3 kg/ha). It has been highlighted that K is the most abundant element in the soils' mineralogy (Table 8) and this explains the high losses. The same trend that was established for N and P was also found with K, where more K was lost with bare fallow (40 kg/ha) and conventional tillage (25 kg/ha) as compared to the conservation tillage systems (0.6 and 4 kg/ha for mulch ripping and tied ridging respectively), see Table 11. The overall treatment differences were significant at P < 0.001 mainly, due to significantly higher soil losses between the treatments. The different years also gave rise to different K losses, which were significant at P < 0.001. These differences show the conservation merits of the conservation tillage treatments, implying that potassium is also conserved effectively through the ability of these treatments in reducing erosion.

Treat/Year (Rainfall)	Year 1 (483 mm)	Year 2 (384 mm)	Overall mean	Source of variation	K loss with erosion
CT	42.3	6.7	24.5	Treat	***
MR	1.0	0.2	0.6	Year	***
TR	8.3	0.2	4.3	Treat x Year	***
BF	66.7	12.9	39.8	MR vs TR	NS
Overall mean	29.6	5.0	17.3	CT vs (MR, TR)	***
n = 6 (Treatment)	s.e.d. = 5.49	s^2 = 90.44			
n = 12 (Year)	s.e.d. = 3.88	df = 16			
n = 3 (Treatment x Year)	s.e.d. = 7.76				

Key: CT = Conventional Tillage; MR = Mulch Ripping; TR = Tied Ridging; BF = Bare Fallow

Table 11. Total potassium loss (kg/ha) as a result of erosion under different tillage systems over three years at MakoholiContill site

Overall the enrichment ratios (soil nutrient concentration: sediment nutrient concentration) for the different nutrients were not very different from one another (Table 12). These were as follows: N: 4.3; P: 3.8 and K: 4.2. Although the amount of P lost with erosion was only a fraction of N and K amounts, it is clear that relative to the amount of P in the soil, all nutrients were lost in near equal proportions. The highest enrichment ratios were recorded under the conservation tillage systems, where the ratios ranged between 6.0 (P) and 7.3 (K), while under conventional tillage the sediments were enriched as follows: 2.0 for N, 1.9 for P and K. The bare fallow recorded the least nutrient enrichment ratios of about 1.0 N and K, while a ratio of 2.7 was recorded for P. The difference in enrichment ratios was only recorded for the different tillage systems and not for the plant nutrients, as these showed a similar trend within these tillage systems.

3.5.2. *Nutrient losses with run-off*

Nitrogen

The amount of nitrogen lost with run-off was very small, on average constituting less than 1% of total nitrogen lost under conventional tillage, bare fallow and tied ridging, while under mulch ripping an average of 2% was recorded over the three years, see Figure 3a. Tied ridging recorded the least N loss of 15 g/ha and conventional tillage the highest N loss of 64 g/ha, however, there was no significant difference between treatments (P = 0.076), see Table 13. A significant difference of P < 0.001 was found between the different years showing that the different rainfall regimes influence run-off amount and consequently nitrogen loss. As N loss with run-off is dissolved N, it is expected that this fraction would be more under cropped treatments where N fertilizer was applied and generally where the nutrient status in the soil is higher. The lower N loss under the bare fallow compared to conventional tillage and mulch ripping, despite higher run-off, is because of this fact. The reason for the low N concentration under the tied ridging treatment is mainly due to the fact that fertilizers are protected on the ridges, while run-off mainly takes place in the furrows.

Treat	**Nutrient concentration in the sediments**			**Enrichment ratios: nutrient in soil: nutrient in sediments**		
	N %	P ppm	K ppm	N	P	K
Year 1						
BF	0.05	39.8	803.9	1.3	1.0	1.5
CT	0.07	104.6	1351.1	1.4	2.0	2.2
MR	0.41	570.9	5397.7	8.2	9.2	9.39
TR	0.28	447.6	5110.6	5.7	4.9	10.5
Year 2						
BF	0.03	14.4	318.7	1.0	0.8	0.6
CT	0.12	61.9	961.5	3.0	2.5	1.6
MR	0.40	156.6	1875.0	8.0	5.2	4.0
TR	0.25	104.8	1813.5	5.06	2.5	4.0
Year 3						
BF	0.05	15.0	-	1.7	0.9	-
CT	0.06	33.4	-	1.5	1.2	-
MR	0.22	124.4	-	4.3	4.8	-
TR	0.52	326.1	-	10.3	10.3	-
-	= missing data					

Key: CT = Conventional Tillage; MR = Mulch Ripping; TR = Tied Ridging; BF = Bare Fallow

Table 12. Nutrient concentrations in the sediments and enrichment ratios for different tillage systems at MakoholiContill site

Phosphorus

The amount of P lost with run-off constituted a slightly higher percentage of total P loss with sediments than was the case with nitrogen. The conservation tillage treatments realized a higher ratio of dissolved P losses, averaging 4% under mulch ripping and 2% under tied ridging. An average of 1% was recorded under conventional tillage and the lowest percentage loss was found under the bare fallow, where P in run-off only constituted 0.6% of total P lost, (Figure 3b). As was the case with N, there were no significant differences between treatments as the amounts were generally very low (Table 14). However, the different years gave rise to significantly different P losses (P < 0.001), due to the different amounts of run-off realized during these years.

Treat/Year (Rainfall)	Year 1 (483 mm)	Year 2 (384 mm)	Year 3 (765 mm)	Overall mean	Source of variation	Nitrogen in run-off
CT	0.0469	0.0071	0.1377	0.0639	Treat	NS
MR	0.0042	0.0037	0.1789	0.0623	Year	***
TR	0.0030	0.0012	0.0421	0.0154	Treat x Year	NS
BF	0.0298	0.0058	0.1208	0.0522	MR vs TR	NS
Overall mean	0.0210	0.0045	0.1199	0.0484	CT vs (MR, TR)	NS
n = 9 (Treatment)	s.e.d. = 0.01986	$s^2 = 0.001775$				
n = 12 (Year)	s.e.d. = 0.01720	df = 24				
n = 3 (Treatment x Year)	s.e.d. = 0.03440					

Key: CT = Conventional Tillage; MR = Mulch Ripping; TR = Tied Ridging; BF = Bare Fallow

Table 13. Nitrogen loss (kg/ha) with run-off under different tillage systems over three years at MakoholiContill site

Treat/Year (Rainfall)	Year 1 (483 mm)	Year 2 (384 mm)	Year 3 (765 mm)	Overall mean	Source of variation	P with run-off
CT	0.0066	0.0028	0.0153	0.0083	Treat	NS
MR	0.0021	0.0003	0.0078	0.0034	Year	***
TR	0.0034	0.0004	0.0081	0.0040	Treat x Year	NS
BF	0.0041	0.0026	0.0089	0.0052	MR vs TR	NS
Overall mean	0.0040	0.0015	0.0100	0.0052	CT vs (MR, TR)	NS
n = 9 (Treatment)	s.e.d. = 0.000973	$s^2 = 0.00004259$				
n = 12 (Year)	s.e.d. = 0.000843	df = 24				
n = 3 (Treatment x Year)	s.e.d. = 0.001685					

Key: CT = Conventional Tillage; MR = Mulch Ripping; TR = Tied Ridging; BF = Bare Fallow

Table 14. Phosphorus loss (kg/ha) with run-off under different tillage systems over three years at MakoholiContill site

Potassium

Dissolved potassium loss with run-off, as was the case with the other elements, constituted a smaller percentage of the total K lost with erosion. The highest percentage was found under the mulch ripping treatment (15 - 20% of total K), followed by tied ridging (5%), then conventional tillage with 2 - 3% and the bare fallow had the least percentage averaging 1% of total K lost, (Figure 3c). The treatments however, did not differ significantly from one another but the different years differed significantly at P < 0.001 (Table 15). Unlike the other elements the loss of dissolved K was highest under the bare fallow, indicating that K is abundant in the soil and highly soluble in water. This also gives an indication on the availability of K in these soils.

Treat/Year (Rainfall)	**Year 1 (483 mm)**	**Year 2 (384 mm)**	**Year 3 (765 mm)**	**Overall mean**	**Source of variation**	**K in run-off**
CT	0.87	0.18	3.81	1.62	Treat	NS
MR	0.20	0.03	3.25	1.16	Year	***
TR	0.38	0.01	1.85	0.74	Treat x Year	NS
BF	0.91	0.02	6.19	2.38	MR vs TR	NS
Overall mean	0.59	0.06	3.77	1.47	CT vs (MR, TR)	NS
n = 9 (Treatment)	s.e.d. = 0.663	$s^2 = 1.976$				
n = 12 (Year)	s.e.d. = 0.574	df = 24				
n = 3 (Treatment x Year)	s.e.d. = 1.148					

Key: CT = Conventional Tillage; MR = Mulch Ripping; TR = Tied Ridging; BF = Bare Fallow

Table 15. Dissolved K loss (kg/ha) with run-off under different tillage systems over three years at MakoholiContill site

3.5.3. Nutrient losses with suspended material

Nitrogen

For all the cropped treatments, most of the N was lost with suspended material and ranged from 49 - 82% of total nitrogen loss under conventional tillage; 86 - 99% under mulch ripping and 93 - 97% under tied ridging. The percentage was lower under the bare fallow and ranged from 29 - 50%, due to the extra-ordinarily high losses of sludge as compared to suspended material (Figure 3a). Analysis of variance showed that nitrogen loss with suspended material differed significantly (P < 0.001) between the different treatments, as a result of the significant treatment differences in the loss of suspended material (Table 16). The conservation tillage treatments did not differ significantly from each other. The different years also gave significant differences in nitrogen loss (P < 0.001). This finding indicates that the suspended material is by far the most important medium for overland transport of nitrogen from arable lands, as a result of erosion. The nitrogen concentration in suspended material ranged from 0.2 - 0.65% compared to 0.04 – 0.05% in the soil.

Phosphorus

Most of the P was also lost with suspended material under the cropped treatments (Table 17). Conventional tillage lost 62 - 83% of total P with suspended material, while the losses ranged between 93 and 97% under mulch ripping and between 91 and 97% under tied ridging. Under the bare fallow, this phenomenon was less pronounced, with the P lost with this fraction accounting for 27 - 61% of total P lost (Figure 3b).

Treat/Year (Rainfall)	Year 1 (483 mm)	Year 2 (384 mm)	Year 3 (765 mm)	Overall mean	Source of variation	Nitrogen in susp.
CT	10.16	5.62	11.38	9.05	Treat	***
MR	0.52	0.16	5.23	1.97	Year	***
TR	2.94	0.14	4.68	2.59	Treat x Year	***
BF	12.33	2.66	22.03	12.34	MR vs TR	NS
Overall mean	6.49	2.14	10.83	6.49	CT vs (MR, TR)	***
n = 9 (Treatment)	s.e.d. = 1.307	$s^2 = 7.686$				
n = 12 (Year)	s.e.d. = 1.132	df = 24				
n = 3 (Treatment x Year)	s.e.d. = 2.264					

Key: CT = Conventional Tillage; MR = Mulch Ripping; TR = Tied Ridging; BF = Bare Fallow

Table 16. Nitrogen loss (kg/ha) with suspended material under different tillage systems over three years at MakoholiContill site

Treat/Year (Rainfall)	Year 1 (483 mm)	Year 2 (384 mm)	Year 3 (765 mm)	Overall mean	Source of variation	P in susp.
CT	1.0541	0.1512	0.4157	0.540	Treat	***
MR	0.0549	0.0087	0.1939	0.086	Year	***
TR	0.2739	0.0073	0.2049	0.162	Treat x Year	***
BF	0.7739	0.0664	0.5244	0.455	MR vs TR	NS
Overall mean	0.539	0.058	0.335	0.311	CT vs (MR, TR)	***
n = 9 (Treatment)	s.e.d. = 0.0597	$s^2 = 0.01602$)				
n = 12 (Year)	s.e.d. = 0.0517	df = 24				
n = 3 (Treatment x Year)	s.e.d. = 0.1033					

Key: CT = Conventional Tillage; MR = Mulch Ripping; TR = Tied Ridging; BF = Bare Fallow

Table 17. Phosphorus loss (kg/ha) with suspended material under different tillage systems over three years at MakoholiContill site

Due to this variation in the different treatments, analysis of variance among the different treatments gave a significant difference at P < 0.001. The different years and the interaction between treatment and year were also significantly different. The finding also shows the significance of suspended material in transporting P from arable lands as a result of erosion.

Potassium

The suspended material accounted for most of the potassium losses under all the cropped treatments, see Table 18. The conservation tillage treatments realized the highest percentages that ranged between 80 and 85% for mulch ripping and 90 and 93% for tied ridging, while conventional tillage lost 66 - 79% of total potassium with this sediment fraction. The bare fallow was the only exception, with the losses as low as 28 - 51% (Figure 3c). Once again this is an indication of the ratio between suspended material and coarse material under the bare fallow. The analysis of variance gave highly significant differences (P < 0.001) between treatments and years and a significant difference of P < 0.01 for the treatment x year interaction. Although somewhat lower the relationship between K and fine soil particles is clearly indicated.

Treat/Year (Rainfall)	**Year 1 (483 mm)**	**Year 2 (384 mm)**	**Overall mean**	**Source of variation**	**K in susp.**
CT	28.1	5.3	16.7	Treat	***
MR	0.8	0.2	0.5	Year	***
TR	7.7	0.2	4.0	Treat x Year	**
BF	34.0	3.6	18.8	MR vs TR	NS
Overall mean	17.7	2.3	10.0	CT vs (MR, TR)	***
n = 6 (Treatment)	s.e.d. = 3.60	s^2 = 38.77			
n = 12 (Year)	s.e.d. = 2.54	df = 16			
n = 3 (Treatment x Year)	s.e.d. = 5.08				

Key: CT = Conventional Tillage; MR = Mulch Ripping; TR = Tied Ridging; BF = Bare Fallow

Table 18. Potassium loss (kg/ha) with suspended material under different tillage systems over three years at MakoholiContill site

The nutrient enrichment ratios for suspended material were generally higher than those recorded for the total sediments, due to the high proportion of fine soil particles (Table 19). The following overall enrichment ratios were found: N: 6.2; P: 5.9 and K: 6.8. Once again the enrichment ratios show that nutrients were lost, relative to their nutrient status in the soil. Although the conservation tillage systems generally had high enrichment ratios, the difference between the treatments was not as distinct as was the case with total sediments. This is as a result of the similar composition of the suspended material, regardless of tillage treatment. The nutrient enrichment ratios were similar across all the nutrients.

http://dx.doi.org/10.5772/3402

3.5.4. Nutrient losses with sludge

Nitrogen

Under all the cropped treatments, the amount of nitrogen lost with sludge was significantly lower than that lost with suspended load (Figure 3a). This phenomenon was more pronounced under the conservation tillage treatments. Conventional tillage recorded 18 - 50%, while mulch ripping recorded 0 - 11% and tied ridging 3 - 7 % of total N loss with sludge. The bare fallow recorded more N loss with sludge (50 - 71%) due to the very high sludge loss. These findings indicate that less nitrogen is associated with coarse soil particles and this is further implicated by the less nitrogen concentration in sludge, ranging between 0.00 and 0.05% as compared to that in suspended material (0.2 - 0.65%). The amount of N lost with sludge differed highly significantly ($P < 0.001$) between the different treatments and it differed significantly at $P < 0.01$ among the different years. As expected, there was no significant difference between the two conservation tillage treatments (Table 20).

Treat	**Nutrient concentration in suspension**			**Enrichment ratios: nutrient in soil: nutrient in suspension**		
	N %	P ppm	K ppm	N	P	K
Year 1						
BF	0.18	267.6	4634.8	4.5	6.8	8.4
CT	0.24	426.2	5233.2	4.8	8.2	8.5
MR	0.41	570.9	5397.7	8.2	9.2	9.4
TR	0.37	584.2	6660.3	7.4	6.4	13.7
Year 2						
BF	0.18	73.1	1671.2	6.0	4.3	3.0
CT	0.26	129.5	1922.5	6.5	5.2	3.1
MR	0.40	156.6	1875.0	8.0	5.3	3.3
TR	0.35	142.8	2505.6	7.0	3.4	5.1
Year 3						-
BF	0.16	73.5	-	5.3	4.6	-
CT	0.21	125.3	-	5.3	4.6	-
MR	0.32	192.2	-	6.4	7.4	-
TR	0.27	163.6	-	5.4	5.1	-

Key: CT = Conventional Tillage; MR = Mulch Ripping; TR = Tied Ridging; BF = Bare Fallow

Table 19. Nutrient concentration in the soil (0 - 25 cm) versus nutrient concentration in suspended material and enrichment ratios for different tillage systems at MakoholiContill site

Treat/Year (Rainfall)	Year 1 (483 mm)	Year 2 (384 mm)	Year 3 (765 mm)	Overall mean	Source of variation	N in sludge
CT	7.19	1.20	11.70	6.70	Treat	***
MR	0.00	0.00	0.65	0.22	Year	**
TR	0.09	0.01	0.20	0.10	Treat x Year	*
BF	19.75	6.39	21.95	16.03	MR vs TR	NS
Overall mean	6.76	1.90	8.62	5.76	CT vs (MR, TR)	***
n = 9 (Treatment)	s.e.d. = 1.962	s^2 = 17.32				
n = 12 (Year)	s.e.d. = 1.699	df = 24				
n = 3 (Treatment x Year)	s.e.d. = 3.398					

Key: CT = Conventional Tillage; MR = Mulch Ripping; TR = Tied Ridging; BF = Bare Fallow

Table 20. Nitrogen loss (kg/ha) with sludge under different tillage systems over three years at MakoholiContill site

Treat/Year (Rainfall)	Year 1 (483 mm)	Year 2 (384 mm)	Year 3 (765 mm)	Overall mean	Source of variation	P in sludge
CT	0.3423	0.0280	0.2350	0.4009	Treat	***
MR	0.0000	0.0000	0.0063	0.0033	Year	***
TR	0.0047	0.0003	0.0050	0.0021	Treat x Year	***
BF	0.4910	0.1760	0.5357	0.2018	MR vs TR	NS
Overall mean	0.2095	0.0511	0.1955	0.1520	CT vs (MR, TR)	***
n = 9 (Treatment)	s.e.d. = 0.02531	s^2 = 0.002883				
n = 12 (Year)	s.e.d. = 0.02192	df = 24				
n= 3 (Treatment x Year)	s.e.d. = 0.04384					

Key: CT = Conventional Tillage; MR = Mulch Ripping; TR = Tied Ridging; BF = Bare Fallow

Table 21. Phosphorus loss (kg/ha) with sludge under different tillage systems over three years at MakoholiContill site

Phosphorus

The amount of P lost with sludge was, as expected, lower than that lost with suspended material for all the cropped treatments (Figure 3b). During the two out of three seasons, there was no phosphorus loss with sludge under mulch ripping and during the last year, the P lost with this sediment fraction constituted only 3% of total P lost. Under tied ridging the losses were nearly the same, ranging from 2 - 4%. The conventional tillage treatment realized significantly higher losses between 16 and 35% of total P lost and once again the bare fallow recorded, during two of the three years, more than 50% of total P lost. The overall treatment and year differences were significant at $P < 0.001$ (Table 21). It is clear once again that P is associated with fine soil particles and not with the non-reactive coarse material as is evidenced by the low P concentrations in sludge, ranging from 0 - 34 ppm compared to 73 - 584 ppm in suspended material.

Potassium

The sludge fraction constituted the lowest losses of K under the cropped treatments, ranging from 0 - 5% under the conservation tillage treatments, a maximum of 32% under conventional tillage (Figure 3c). Under the bare fallow 48 - 72 % of total K was lost with this sediment fraction. These obvious differences between both treatment and year factors were significant at $P < 0.001$ (Table 22). K is therefore associated with the suspended material than with sludge, the somewhat higher percentages lost under conventional tillage and bare fallow are merely in relation to the very high coarse material lost under these treatments as the actual nutrient concentration in the sludge is very low compared to that in suspended material, a range of 0 - 472 ppm in sludge and 1671 - 6660 ppm in suspended material.

Treat/Year (Rainfall)	**Year 1 (483 mm)**	**Year 2 (384 mm)**	**Overall mean**	**Source of variation**	**K in sludge**
CT	13.35	1.17	7.26	Treat	***
MR	0.00	0.00	0.00	Year	***
TR	0.17	0.01	0.09	Treat x Year	***
BF	31.75	9.32	20.54	MR vs TR	NS
Overall mean	11.32	2.62	6.97	CT vs (MR, TR)	***
n = 6 (Treatment)	s.e.d. = 2.126	$s^2 = 13.56$			
n = 12 (Year)	s.e.d. = 1.503	df = 16			
n = 3 (Treatment x Year)	s.e.d. = 3.006				

Key: CT = Conventional Tillage; MR = Mulch Ripping; TR = Tied Ridging; BF = Bare Fallow

Table 22. Potassium loss (kg/ha) with sludge under different tillage systems over three years at MakoholiContill site

Due to the high proportion of non-reactive coarse particles in sludge, the nutrients concentration was low compared to total sediments and suspended material. Generally all the nutrients under all tillage systems recorded lower nutrient concentrations in the sludge compared to nutrient concentrations in the original soil, resulting in enrichment ratios less than 1.0, with the exception of N under bare fallow, which recorded 1.0. The overall nutrient enrichment ratios in sludge were as follows: N: 0.6; P: 0.5 and K: 0.5, an indication that this sediment fraction is impoverished in plant nutrients compared to the original soil (Table 23). Furthermore, there is no report on the association of coarse soil particles and the fertility of a soil, as is the case with fine soil particles. Generally, the sandier the soil the lower its nutrient status and/or soil productivity.

Treat	Nutrient concentration in sludge			Enrichment ratios: nutrient in soil: nutrient in sludge		
	N %	P ppm	K ppm	N	P	K
Year 1						
BF	0.04	17.5	429.8	1.0	0.4	0.8
CT	0.03	31.8	472.8	0.6	0.6	0.8
MR	0.00	0.0	0.0	0.0	0.0	0.0
TR	0.02	33.6	414.7	0.4	0.4	0.9
Year 2						
BF	0.02	11.0	240.8	0.7	0.6	0.4
CT	0.03	17.7	333.3	0.8	0.7	0.5
MR	0.00	0.0	0.0	0.0	0.0	0.0
TR	0.01	10.0	83.3	0.2	0.2	0.2
Year 3						
BF	0.04	9.3	-	1.3	0.6	-
CT	0.03	16.4	-	0.8	0.6	-
MR	0.05	13.2	-	1.0	0.5	-
TR	0.04	26.0	-	0.8	0.8	-

Key: CT = Conventional Tillage; MR = Mulch Ripping; TR = Tied Ridging; BF = Bare Fallow

Table 23. Nutrient concentrations in the soil (0 - 25 cm) versus nutrient concentrations in the sludge and enrichment ratios for different tillage systems at MakoholiContill site

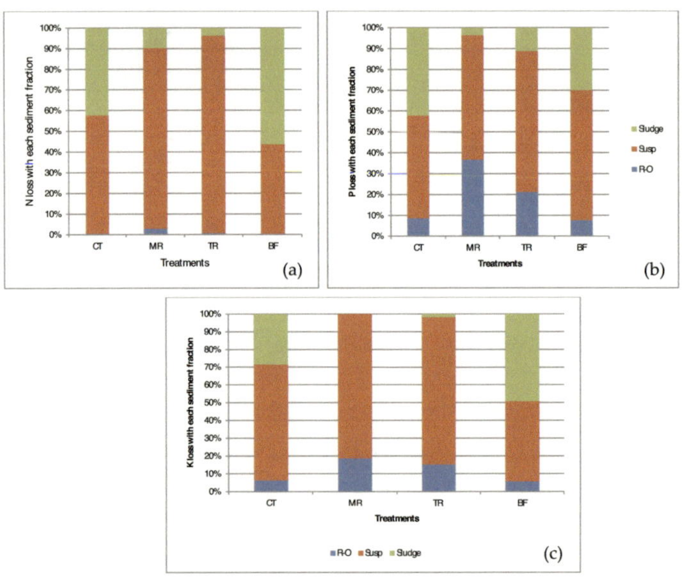

Key: R-O = Run-off; Susp = Suspended material

Figure 3. (a) Nitrogen, (b) phosphorus and (c) potassium losses as influenced by different erosion fractions under four tillage systems at MakoholiContill site (averages over three years)

3.5.5. Eroded nutrients versus soil loss and sediment fraction

Regression analysis was carried out to relate nutrient loss with the amount of soil lost and sediment fraction. Firstly, a general regression analysis was carried out, where all the data collected was pooled, i.e. without specifying the treatments or the years and soil loss, suspended material and sludge were considered independently (Table 24). Data was then split according to the different treatments (disregarding years) and again the different elements were regressed with soil loss and sediment fractions. From the regression output, each element was then calculated in relation to a tonne of lost soil and/ or sediment fraction. Correlation coefficients were also worked out for the relationship between each element and soil loss as well as sediment fraction (Tables 24 and 25).

Treat/Year	**Element**	**Element kg/1t SL**	**Standard error**	**% variance accounted for**	**P value**	**Correlation SL:Element**
Pooled	N	0.360	0.019700	94.5	***	0.980
Pooled	P	0.010	0.002090	38.3	***	0.719
Pooled	K	0.767	0.104000	80.0	***	0.908
Treat/Year	Element	Element kg/1tSusp.	Standard error	% variance accounted for	P value	Correlation Susp:Element
Pooled	N	1.589	0.0416	95.4	***	0.977
Pooled	P	0.058	0.00722	40.6	***	0.654
Pooled	K	4.201	0.271	86.5	***	0.932
Treat/Year	Element	Element kg/1t Sludge	Standard error	% variance accounted for	P value	Correlation Sludge:Element
Pooled	N	0.186	0.0137	76.5	***	0.879
Pooled	P	0.005	0.000302	80.0	***	0.904
Pooled	K	0.390	0.0198	92.1	***	0.960

SL = Soil loss; Susp = suspended material

Table 24. Nutrient loss as affected by soil loss, sludge and suspended material over three seasons at MakoholiContill site

The results of the regression analysis show that pooling the data gave moderate nutrient losses for every tonne of soil lost. All the nutrients were below 1 kg for every 1 tonne of soil lost, i.e., total sediments (ranging from 0.01 for P to 0.7 kg for K). The amounts of the nutrient losses were related to the losses under bare fallow but these amounts would under estimate the losses under cropped treatments. Generally for the pooled estimates, K was the most abundant element in the sediments and the sequence could be summed up as follows: K > N > P. The variance accounted for in the estimates was also very high for N and K and low for P.

The sediment composition also influenced the amount of nutrients per unit of soil loss, with more nutrients lost with suspended material than with coarse material (Table 24). This table shows that an average of 1.589 kg N was lost with one tonne of suspended material compared to 0.186 kg N lost with one tonne of sludge, i.e. (8.5 times). About 12 times more P was lost with one tonne of suspended material than with sludge, while K was 11 times more in suspended material than in sludge. This information further consolidates the fact that much more nutrients are lost with suspended material regardless of tillage treatment and plant element. The loss of coarse soil particles should have implications on soil productivity mainly due to the reduction of soil tilth and not soil fertility.

The different treatments also showed that the conservation tillage treatments lost more nutrients per unit soil loss than conventional tillage systems (Table 25), due to the low sludge: suspension ratio in the former. For the same reason, conventional tillage also lost more nutrients (all elements) per tonne of soil loss than the bare fallow. Between the two conservation tillage treatments, more nutrients (N, P and K) were lost under tied ridging than under mulch ripping. The differences though, were not significant. All the treatments showed a similar trend as that of pooled data. P losses were highly correlated to soil loss under mulch ripping, followed by bare fallow, whereas under conventional tillage and tied ridging the correlation was rather low, although still significant. The poor correlation may be as a result of the very low P losses, which may affect the accuracy of such measurements.

Treat.	Element	Element kg/1t SL	Standard error	% variance accounted for	P value	Correlation SL:Element
BF	N	0.305	0.030000	70.9	***	0.842
BF	P	0.008	0.001270	29.9	***	0.614
BF	K	0.700	0.105000	72.0	***	0.958
CT	N	0.434	0.044200	54.2	***	0.891
CT	P	0.017	0.005070	very low	**	0.339
CT	K	1.199	0.073400	95.1	***	0.977
MR	N	1.242	0.041400	98.7	***	0.994
MR	P	0.028	0.002420	89.5	***	0.966
MR	K	4.600	0.659000	80.1	***	0.951
TR	N	1.437	0.150000	79.0	***	0.900
TR	P	0.059	0.016700	11.7	*	0.496
TR	K	5.155	0.359000	95.7	***	0.981

SL = Soil loss; Susp = suspended material; CT = Conventional Tillage; MR = Mulch Ripping; TR = Tied Ridging; BF = Bare Fallow

Table 25. The relationship between nutrient loss and soil loss under different tillage systems at MakoholiContill site

There is evidence that a substantial amount of nutrients is lost with erosion, as shown by the overall averages of 12.3 kg/ha N; 0.5 kg/ha P and 17.3 kg/ha K. The amount of nutrient lost was found to be strongly dependent on the nutrient status of the soil, i.e. the higher the status of a particular nutrient in the soil, the higher its loss with erosion. The nutrient status of the soils showed the following trend K > N > P and the overall nutrient loss with erosion also showed exactly the same trend. This explains why soils with higher fertility status lose much more nutrients relative to those with a lower fertility status (Stoorvogel and Smaling, 1990). According to Rose *et al.* (1988), the amount of a nutrient lost with erosion is dependent upon the soil type, tillage practice and the type of erosion. From this study it was found that the amount of soil loss and the sediment fraction – including run-off – were also important in determining the amount of nutrient loss, especially on sandy soils, where the amount of clay and organic matter are critical as sources of plant nutrients.

The sediment fraction on sandy soils is very important in determining the amount of nutrient loss due to the selective nature of sheet erosion on these soils. Nutrient losses in the water portion of the run-off were small, almost negligible compared to the losses with solid sediments, ranging from 1 – 2 % of total N; 0.6 – 4% of total P and 1 – 20% of total K. This was expected, as these nutrients have to be dissolved in water. Even in the original soil, the nutrients in the soil solution are only a small fraction of nutrients sorbed in the soil, ranging from 0.001% for P to 25% for Ca (Brady, 1984; Stevenson, 1985; Singer and Munns, 1987). The solid fraction is therefore, the major source of plant nutrient loss (Barisas et al., 1978; Kejela, 1991). Suspended material is the main source of nutrient loss from agricultural lands, as evidenced by the very high percentages of nutrient losses with this sediment fraction, especially under conservation tillage, > 90%. Although the ratio between sludge and suspension under the conventional tillage was between 1.5 and 5, about 25% of total sediments, it accounted for 63% of total N, 74% of total P and 73% of total K lost with erosion. With a sludge suspension ratio of between 10 and 17, the bare fallow also recorded proportionally higher nutrients with suspended than with sludge, shown by the following percentages: 39% of total N and K and 46% of total P being lost with this 8% suspended material. This finding certainly shows that much more nutrients are lost with suspended material than with any other sediment fraction.

Furthermore, the N concentration in sludge ranged from 0 – 0.5%, while in suspended material it ranged from 0.2 – 0.65 %, P concentration in sludge ranged from 0 – 34 ppm compared to 73 – 584 ppm and K recorded a concentration of 0 – 472 ppm in sludge and 1671 – 6660 ppm in suspended material. This is because clay and organic matter are the sorption sites for much of the nutrients and organic matter is also crucial in the cycle of P and N. Brady (1984) reported that organic matter was the major indigenous source of N while 65% of total P in the soil was found in the form of organic compounds. Clay more than organic matter, is the main source of fixed K and other cations and K losses are therefore associated with clay loss. Due to the selective nature of sheet erosion, high affinity of P to adsorption, fixation of K and ammonium ions, as well as the presence of K ions in clay minerals, erosion is the main source of nutrient and productivity loss in agricultural lands.

The affinity of the nutrients to the fine soil particles cannot be doubted. The exchange sites on the clay minerals and organic matter are the basis for this affinity, as nutrients are held at these exchange sites (Brady, 1984; Stevenson, 1985). According to Singer and Munns (1987), the different clay minerals and humus are most important in holding nutrients due to their specialized surface properties, capable of chemically retaining individual nutrients. Tiessen, Cuevas and Salcedo (1998) and Stocking (1984) also reported that soil organic matter provided plant nutrients in low-input agriculture and that N and P release depended on the mineralization of organic matter while cation exchange depended on the maintenance of organic matter. This is why the loss of top soil is detrimental to any soils' productivity as there is a close association between clay, organic and the plant nutrients. The proximity and concentration of organic matter near the soil surface and close association with plant nutrients, make the erosion of soil organic matter a strong indicator of overall plant nutrients resulting from erosion (Follet*et al.*, 1987).

The results across all cropped treatments and for all the elements show that most of the nutrients lost with erosion are associated with suspended material. Under conservation tillage systems it is arguable that the high percentages may be due to the fact that most of the soil lost was in suspended form, however, the high losses attributed to this fraction under conventional tillage indicate otherwise. This finding proves beyond doubt that although less suspended material may be lost from a field, it carries most of the soil nutrients with it. The conservation tillage systems obviously have higher percentages of nutrient losses with suspended material, however the quantities of nutrients lost are negligible when compared to those lost under conventional tillage and bare fallow because of reduced soil losses under conservation tillage.

The nutrient losses with sludge are minimal when compared to those lost with suspended material. While as high as 92% of total soil loss under bare fallow is sludge, the percentages of nutrients lost with this fraction are not as high (56% N; 23% P and 52% K). It should be noted also that there was no distinct separation of sludge and suspension, due to the fact that some suspended material would settle with the sludge, during a storm, before sampling was carried out. This explains the presence of 0.81 - 4.02% clay and 0.04 - 0.55% organic matter in the sludge. The nutrients in the sludge can be attributed to the fine soil particles and not the coarse material.

Nutrient losses (N, P and K) varied significantly among the different treatments. The conservation tillage treatments lost significantly less nutrients compared to the conventional tillage systems. Here, 2.3 and 2.7 kg/ha N; 0.09 and 0.2 kg/ha P; 0.6 and 4.3 kg/ha K were lost under mulch ripping and tied ridging respectively compared to 15.8 and 28.4 kg/ha N; 0.8 and 0.9 kg/ha P; 24.5 and 39.8 kg/ha K under conventional tillage and bare fallow respectively. As these treatments lost significantly different amounts of sediments, also following the same trend, this indicates that the nutrient losses with erosion are closely associated with the rate of soil loss (Elwell and Stocking 1988; Kejela 1991). The tillage systems in this study also showed their effect on the amount of nutrients lost by determining the amount of soil loss. Due to the fact that plant nutrients sorbed to the soil are transported with eroding sediments, the amount of soil lost with erosion becomes very important in determining the amount of nutrients lost. The conservation tillage systems dramatically reduced losses of soil and total nutrients when compared to conventional tillage systems, however the nutrient concentrations per unit soil loss are higher than for conventional tillage systems.

The concentration of nutrients in the sediments was much higher under the conservation tillage systems as compared to conventional tillage, obviously as a result of a high percentage of fine particles in the sediments compared to the later. Very high enrichment ratios of all nutrients were thus recorded in the sediments of the conservation tillage systems as a result of the high affinity of nutrients to fine soil particles (Barisas*et al.* 1978). However, the advantage of low amount of sediments in conservation tillage also resulted in lower average losses under this system.

The different years lost significantly different amounts of nutrients, which depended on the amount of rainfall received and amount of soil lost. For all the nutrients, nutrient losses increased with the increase in rainfall amount, i.e. with increased sediments. The regression analysis that was carried out to find the relationship between soil loss ad nutrient loss showed that nutrient losses are highly dependent on soil losses and that if soil losses are known, nutrient losses can be confidently predicted. The conservation tillage systems lose more nutrients per unit soil loss than conventional tillage systems because their sediments are predominantly fine particles, e.g., per tonne of soil lost 1.4 and 1.2 kg/ha of N were predicted to be lost under TR and MR respectively, while BF and CT were predicted to lose 0.3 and 0.4 kg/ha respectively. The point on the high affinity of all the nutrient elements to the fine soil particles has also been emphasized.

By conserving the soil, nutrients are conserved and nutrient replacement costs of erosion are drastically reduced, especially if the value of sustainable production is also taken into consideration. It should be emphasized, however, that the loss of organic matter and clay and resultant physical degradation of the soil, leading to poor tilth, low available water holding capacity and high bulk density, was not evaluated. This means that the value of nutrient losses is but a fraction of total loss (Kejela, 1991).

4. Conclusions

Sheet erosion is a selective process that robs the soil of its fine particles, i.e. clay and organic matter. The high enrichment ratios of clay and organic matter found in sediments as compared to the original soil, serve to support this fact. Of the two sediment fractions, the soil lost in suspension is the most detrimental as it comprises of clay and organic matter particles, which are known to be the soils' plant nutrient reservoirs. There is a very high association between nutrients and fine soil particles as shown by the high amount of nutrients lost with a unit mass of suspended material as compared to those lost with the same unit mass of sludge and/ or total sediments. This makes the suspended material the most detrimental sediment fraction, negatively affecting the soils' fertility status as well as impacting negatively on the soil's physical condition. The suspended material recorded high concentrations of clay, organic matter and nutrients when compared to sludge. However, the total loss of clay, organic matter and plant nutrients in the sediments is not dependent upon their concentrations in the eroded soil but rather on the total amount of soil lost. Thus mulch ripping and tied ridging proved to be effective in maintaining clay and organic matter levels and thus significantly reducing nutrient losses from agricultural lands due to their ability to reduce soil erosion.

Acknowledgements

I would like to express my gratitude to GTZ for providing the much needed funding through CONTILL (Conservation Tillage), a collaborative project between GTZ and the Government of Zimbabwe (GoZ). Further acknowledgement goes to the GoZ, for providing me with the opportunity and research facilities. I would like to thank all the CONTILL members from both Domboshawa and especially Makoholi site for their relentless support and input towards the success of this project.

Author details

Adelaide Munodawafa

Midlands State University, Zimbabwe

References

[1] Adams, J. E. (1966). Influence of mulches on run-off, erosion and soil moisture depletion. Soil Science of America Proc, , 30, 110-114.

[2] Angers, D. A., N'dayegamiye, A., & Cote, D. (1993). Tillage induced differences of particle-size fraction and microbial biomass. Journal of Soil Science Society of America, , 57(3), 234-240.

[3] Anon, (1969). Guide to Makoholi Experiment Station. Department of Research and Specialist Services, Salisbury.

[4] Aylen, D. (1939). Soil and Water Conservation, Part 1.The Rhod. Agric. Jour. , 1095, 1-9.

[5] Barisas, S. G., Baker, J. L., Johnson, H. P. ., & Laflen, J. M. (1978). Effect of tillage systems on run-off losses of nutrients, Iowa Agriculture and Home economics Experiment StationReport, 85, New Jersey.

[6] Bauer, A., & Black, A. L. (1994). Quantification of the effect of soil organic matter content on soil productivity. JSSSA , 58, 185-193.

[7] Beare, M. H., Hendrix, P. F. ., & Coleman, D. C. (1994). Water-stable aggregates and organic fractions in conventional tillage andno tillage soils.Journal of Soil Science Society of America , 58, 777-786.

[8] Biot, Y. (1986). Modelling of the on site effect of sheet erosion and rill wash on the productivity of the land: A research design. Discussion paper 192.UEA, Norwich.

[9] Brady, N. C. (1984). The nature and properties of soils. Macmillan Publishing Co., New York.

[10] Braithwaite, P. G. (1976). Conservation tillage- Planting systems.Rhodesian Farmer, , 10, 25-32.

[11] Bremner, J. M. ., & Mulvaney, C. S. (1982). Nitrogen- Total.Methods o. f Soil Analysis, Part Chemical and Microbiological Properties (2). Madison, USA., 2.

[12] Cassel, D. K., Raczkowski, C. W., & Denton, H. P. (1995). Tillage effects on corn production and soil physical conditions. JSSSA , 59, 1436-1443.

[13] Cormack, R. M. M. (1953). Conservation and mechanization aspects of the ley. Paper presented at the Bulawayo Congress of the Association of SWC, July 1953.

[14] Elwell, H. A., & Norton, A. J. (1988). No-till tied ridging. A recommended sustained crop production system. Technical Report 3, IAE, Harare.

[15] Elwell, H. A., & Stocking, M. A. (1984). Rainfall parameters and a cover model to predict run-off and soil loss from grazing trials in the Rhodesian sandveld. Grasslands Society for Southern Africa, , 9, 157-164.

[16] Elwell, H. A., & Stocking, M. A. (1988). Loss of soil nutrients by sheet erosion is a major hidden cost. TheZimbabwe Science News, 22, 7/ , 8, 79-82.

[17] Elwell, H. A. (1975). Conservation implications of recently determined soil formation rates in Rhodesia.Science Forum, , 2, 5-20.

[18] Elwell, H. A. (1987). An assessment of soil erosion in Zimbabwe.Zimbabwe Science News, 19, ¾, , 27-31.

[19] Elwell, H. A. (1993). Feasibility of modelling annual soil loss, run-off and maize yiled for the two research sites, Domboshawa and Makoholi. Projections to other Natural Regions in Zimbabwe.Testing of and contributions to SLEMSA. Unpublished Consultancy Report, AGRITEX/ GTZ Conservation Tillage Project, IAE, Harare.

[20] Follet, R. H., Gupta, S. C., & Hunt, P. G. (1987). Conservation practices: relation to the management of plant nutrients for crop production. Journal of Soil Science Society of America, Special Publication 19.

[21] Frede, H. G., & Gaeth, S. (1995). Soil surface roughness as a result of aggregate size distribution 1. Report: Measuring and evaluation method. Journal of Plant Nutrition and Soil Science , 158, 31-35.

[22] Gee, G. W., & Bauder, J. W. (1986). Texture hydrometer method. Methods of of soil analysis, Part 1: Physical and Mineralogical Methods. 2nd Edition.Ed. A. Klute, ASA & SSSA, Wisconsin, USA.

[23] Gerzabek, M. H., Kirchman, H. ., & Pichlmayer, F. (1995). Response of soil aggregate stability to manure amendments in the Ultuna long-term soil organic matter experiment.Journal of Plant Nutrition and Soil Science, , 158, 257-260.

[24] Gill, W. R., & vanden, Berg. G. E. (1967). Soil dynamics in tillage and traction. US Department of Agriculture/ Agricultural Research Services: Agric Handbook (316), 511.

[25] Godwin, R. J. (1990). Agricultural engineering in development: tillage for crop production in areas of low rainfall. FAO Agricultural Services Bulletin, 83, Rome.

[26] Grant, P. (1981). Peasant farming on infertile sands.Rhod. Sci. News 10 (10), 252-254. Harare, Zimbabwe.

[27] Grant, P. M. (1976). Peasant farming on infertile sands.Rhodesian Science News, , 10(10), 252-254.

[28] Hanotiaux, G. (1980). Run-off, erosion and nutrient losses on loess soils in Belgium.Assessment of Erosion Eds. M. de Broodt and D. Gabriels), Wiley, Chichester, UK., 369-378.

[29] Hudson, N. W. (1959). Results of erosion research in Southern Rhodesia. Advisory Leaflet 13.Conex, Salisbury.

[30] Hudson, N. W. (1992). Land Husbandry. Batsford, London.

[31] Hudson, N. W. (1958). Erosion research.Advisory Notes. Conex, Salisbury.

[32] Hudson, N. W., & Jackson, D. C. (1962). Results achieved in the measurement of erosion and run-off in Southern Rhodesia. Thechn.Memoranda 4, Conex, Salisbury.

[33] Hunt, P. G., Karlen, D. L., Matheny, T. A., & Quisenberry, V. L. (1996). Changes in carbon content of Norfolk loamy sand after fourteen years of conservation and conventional tillage. JSWC , 51(3), 255-258.

[34] Kaihura, F. B. S., Kullaya, I. K., Kilasara, M., Aune, J. B., Singh, B. R., & Lal, R. (1998). Impact of soil erosion on soil productivity and crop yield in Tanzania.Advances in GeoEcology, , 31, 375-381.

[35] Kejela, K. (1991). The cost of soil erosion in Anjeli, Ethiopia.Soil conservation for survival.Proceedings of the 6th International Soil Conservation Organisation, Berne.

[36] Knusden, D., Peterson, G. A., & Pratt, P. F. (1982). Lithium, Sodium and Potassium. Methods of Soil Analysis, Part 2: Chemical and Microbiological Properties. Agronomy 9: 2nd Edition. Ed. A. L. Page, ASA & SSSA, Wisconsin, USA.

[37] Lal, R. (1988). Monitoring soil erosion's impact on crop productivity.Soils Erosion Research Methods.SWCS & ISSS Iowa, USA., 187-202.

[38] Lobb, D. A. (1995). Tillage: Implications for nutrient management. Agricultural Science and Technology Workshop, Truro, Nova Scotia: , 149-154.

[39] Lowery, B., Larson, W. E., & (1995, . (1995). Erosion impact on soil productivity.Journal of Soil Science Society of America, , 59(3), 647-648.

[40] Massey, H. F., & Jackson, M. L. (1952). Selective erosion of soil fertility constituents.Soil Sci. Proc. 16(4).

[41] Ministry of Natural Resources and Tourism (MNRT). (1987). The National Conservation Strategy. The Government of Zimbabwe, Harare.

[42] Moyo, S., Robinson, P., Katerere, Y., Stevenson, S., & Gumbo, D. (1991). Zimbabwe's environmental dilemma.Balancing resource inequities.ZERO. Harare.

[43] Munodawafa, A., & 200, . (2007). Assessing nutrient losses under different tillage systems and their implications on water quality.Journal of Physics and Chemistry of the Earth, , 32, 1135-1140.

[44] Nelson, W. D., & Sommers, L. E. (1982). Total carbon, organic carbon and organic matter.. Methods of Soil Analysis, Part 2: Chemical and Microbiological Properties. Agronomy 9: 2nd Edition. Ed. A. L. Page, ASA & SSSA, Wisconsin, USA.

[45] Nyamapfene, K. (1991). Soils of Zimbabwe.Nehanda Publishers, Harare.

[46] Oliver G.J. and Norton, A.J. (1988). Tillage research and ploughing technology in Zimbabwe.Proceedings of the 11th ISTRO Conference, Edinburg, 11th- 15th July.

[47] Olsen, S. R. ., Sommers, L. E., & Phosphorus, . (1982). Phosphorus.Methods of Soil Analysis, Part 2 - Chemical and M. icrobiological Properties (2). Madison, USA.

[48] Poesen, J., & Savat, J. (1980). Particle size separation during erosion by splash and runoff.Assessment of Erosion 427-440 (eds. M. De Boodt and D. Gabriels) Wiley, Chichester, UK.

[49] Reicosky, D. C., Kemper, W. D., Langdale, G. W., Douglas, C. L., & Rasmussen, P. E. (1996). Soil organic matter changes resulting from tillage and biomass production.*Journal of Soil and Water Conservation* . 50(3), 253-261.

[50] Rose, C. W., Saffigna, P. G., Hairsine, P. B., Palis, R. G., Okwach, G., Proffitt, A. P. B., & Lovell, C. J. (1988). Erosion processes and nutrient loss in land conservation for future generations. Proceedings of the 5th International Soil Conservation Organization, Bangkok, Thailand, January, 1988., 18-29.

[51] Rosenberg, M. (2007). Koeppen climate classification. http://geography.about.com/od/physicalgeography/a/koeppen.htm

[52] Salinas-Garcia, J. R., Hons, F. M. ., & Matocha, J.e. (1997). Long-term effects of tillage and fertilization on soil organic matter dynamics.Journal of Soil Science Society of America , 61, 152-159.

[53] Sauerbeck, D. R. (1984). Soil management, soil functions and soil fertility. Journal of Plant Nutrition and Soil Science , 153(3), 243-248.

[54] Shaxson.T.F. (1975). Soil erosion, water and organic. World crops , 1975, 6-10.

[55] Singer, M. J., & Munns, D. N. (1987). Soils, an introduction. Macmillan Publishing Co. New York.

[56] Stevenson, F. J. (1985). Cycles of soil: Carbon, nitrogen, phosphorus, sulphur, micronutrients. USA.

[57] Stocking, M. (1983). Development projects for the small farmer: Lessons from east and central Africa in adapting soil conservation. Proceedings of the SCSA Conference.

[58] Stocking, M., & Peake, L. (1985). Soil conservation and productivity.Proceedings of IV International Conference on Soil Conservation, November 3-9 1985, 399-438. University of Venezuale. Maracay, Venezuela.

[59] Stocking, M. A. (1984). Soil potentials: an evaluation of a rating method in Zimbabwe. Discussion Paper 172, UEA, Norwich

[60] Stoorvogel, J. J. ., & Smaling, E. M. A. (1990). Assessment of soil nutrient depletion in Sub-Saharan Africa Report 28, Wageningen, The Netherlands, 1983-2000.

[61] Tanaka, D. L. (1995). Spring-wheat straw production and composition as influenced by top soil removal.Journal of Soil Science Society of America, , 59(3), 649-653.

[62] Thompson, J. G., & Purves, W. D. (1978). A guide to the soils of Rhodesia. Department of Research and Specialist Services, Harare

[63] Thompson, J. G., & Purves, W. D. (1981). A guide to the soils of Zimbabwe. Department of Research and Specialist Services, Harare.

[64] Thompson, J. G. (1967). Report on the soils of theMakoholi Experiment Station, Department of Research and Specialist Services, Salisbury.ja

[65] Tiesen, H. E., Cuevas, E. ., & Salcedo, I. H. (1998). Towards sustainable land use, furthering Cooperation between people and institutions.Proceedings of the 13th International Soil Conservation Organisation,Bonn, Germany, August 1996.

[66] Vogel, H. (1992). The effects of conservation tillage on sheet erosion from sandy soils at two experimental sites in Zimbabwe.Applied Geography , 12, 229-242.

[67] Vogel, H. (1993). An evaluation of five tillage systems from small-holder agriculture in Zimbabwe.Der TropenLandwirt, , 94, 21-36.

[68] Wendelaar, F. E. ., & Purkis, A. N. (1979). Recording soil loss and run-off from 300m^2 erosion research field plots.Research Bulletin 24, Conex, Harare.

[69] Willcocks, T. J., & Twomlow, S. J. (1993). A review of tillage methods and soil and water conservation in southern Africa. Soil Tillage Research, , 27, 73-94.

[70] Working Document (1990). Conservation tillage for sustainable crop production systems, IAE, Harare.

[71] Young, K. K. (1980). The impact of erosion in the productivity of the soils in the US. Assessment of Erosion Eds M. de Broodt and D. Gabriels, Wiley, Chichester, UK., 295-304.

Chapter 8

Gully Erosion in Southeastern Nigeria: Role of Soil Properties and Environmental Factors

C.A. Igwe

Additional information is available at the end of the chapter

http://dx.doi.org/10.5772/51020

1. Introduction

The countries of sub-Saharan Africa are besieged by serious environmental degradation resulting in desert encroachment, draught and soil erosion due to either wind impact or very high intensive rainfall resulting in heavy runoff and soil loss. The problems have adversely affected agricultural productivity and thus casting doubt of food security in the zone. The ecological and social settings in the zone are often distorted some times leading to losses in human and material capitals. In Nigeria desertification and aridity are the major environmental problems of the Northern part of the country while the high torrential rainfall of the southern Nigeria creates enabling environment for catastrophic soil erosion in the region.

The greatest threat to the environmental settings of southeastern Nigeria is the gradual but constant dissection of the landscape by soil erosion by water. Although the incipient stages of soil erosion through rill and interrill are common and easily managed by the people through recommended soil conservation practices, the gully forms have assumed a different dimension such that settlements and scarce arable land are threatened. Therefore, gully erosion problems have become a subject of discussion among soil scientists, geographers, geologists, engineers and social scientists. Ofomata [1] indicated that gully erosion types are the most visible forms of erosion in Nigeria mainly because of the remarkable impression they leave on the surface of the earth. Again Ofomata [2] remarked that more than 1.6% of the entire land area of eastern Nigeria is occupied by gullies. This is very significant for an area that has the highest population density 500 persons per km^2 in Nigeria. Before the 1980's the classical gully sites in the region were the Agulu, Nanka, Ozuitem, Oko in Aguata area, Isuikwuato and Orlu. With the increased development activities the number and magnitude escalated thus making many government administrations within the region to set up soil erosion control with different names in different states. At the last count the Federal Government of Nigeria has started showing interest in ecological problems in the country including the control of the gullies which has reached more than 600 active sites in the region. The gullies are also a visible manifestation of the physical loss of the land due to erosion. Long before now a lot of attention has been focused on the control measurers. As early as the 1930s, the colonial government in Nigeria has undertaken a campaign of tree planting with the main objectives of controlling erosion especially on the steep slopes of upland landscapes in the region. Ever since then there has been a constant enquiry as to the causes of these catastrophic erosion. Most researchers [2, 3, 4] have shown that the environmental factors of vegetation, geology, geomorphology, climate in the form of rainfall which is very aggressive in the region and the soil factor all contribute in the erosion problem and their development. The consequence of the soil erosion is loss of land for agriculture and for habitation. During some slides caused by gully formation, lives have been lost while some communities have been separated because of deep and very wide gullies that may reach in some cases 12 m deep and more than 1.5 km long like the Nanka/Agulu gully complexes or in Oko in Aguata, Anambra State. Crop yields have been reduced, thus creating problem in the "green revolution" campaign.

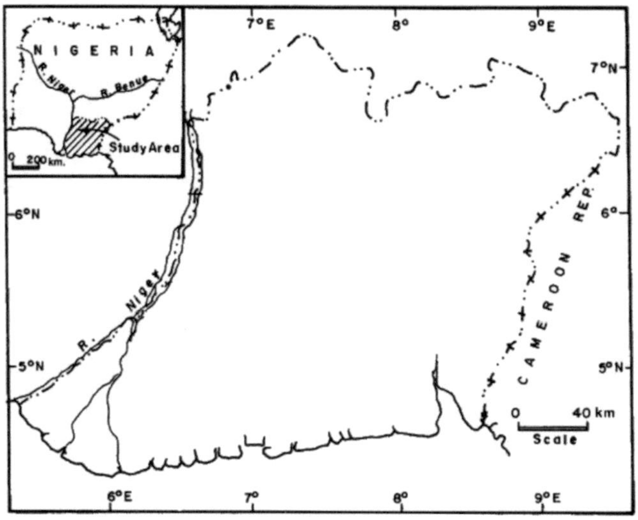

Figure 1. Location map.

Figure 1. Location map of the reviewed area

2. Causes of Soil erosion

Soil erosion generally is caused by several factors working simultaneously or individually to detach, transport and deposit soil particles in a different place other than where they were formed. The resultant effects of this phenomenon are deep cuttings and ravine which dissects the entire land surface. These are very common site all over the geographical region of southeastern Nigeria. It is well established fact among earth scientists that a number of environment factors as well as pedological parameter influence the extent of soil erosion where ever it occurs globally. These factors are perhaps guided by human factors known as anthropogenic factors. Although man has helped in reshaping and preserving the earth surface yet man has also helped in causing instability of equilibrium in the natural ecology and hence the rapid spread of environmental problem such as soil erosion. Igwe [5] noted that the anthropogenic factors are mainly technical factors comprising mainly of land use and tillage methods, the choice and distribution of cultures and the nature of agro-technology. In Northern hemisphere including many countries of Europe, Giordano et al. [6] showed that among the factors that encourage soil erosion are vegetation clearance, intensive harvesting and over-grazing leaving the soil bare. Other factors are soil compaction caused by heavy machinery which reduces the infiltration capacity of the soil and thus promoting excessive water runoff and soil erosion. In classical modelling works on soil erosion prediction and estimation, works by Renard et al. [7], Igwe et al. [8] among others recognised topography/relief, rainfall and soil factors as being the main agents that determine the extent of soil erosion hazard. The soil factor represents the soil erodibility which is also a product of geology and soil characteristics. In showing how these factors influence the extent of soil erosion and gullying in southeastern Nigeria, there is going to be an attempt into discussing how these parameters contribute to gully erosion in other geographical zones.

3. The role of topography

Hudson [9] observed that in simplest terms steep land is more vulnerable to water erosion than flat land for reasons that erosive forces, splash, scour and transport, all have greater effect on steep slopes. Soil erosion generally is a function of slope attributes. The slope length and the amount of soil erosion have always been proportional to the steepness of the slope. Also the slope geometry of hill sides (i.e. whether convex or concave) often contribute significantly to soil loss and gully development. In southeastern Nigeria, Ofomata [3] found that there is a positive relationship between relief and soil erosion while in southwestern Nigeria, Lal [10] observed an increased severity of soil erosion as the slope changed from 5 to 15%. On a 15% slope he recorded a total soil loss of 230 t/ha/yr from bare plots as against soil loss of 11.2 t/ha/yr on 1% slope.

The topography of southeastern Nigeria according to Ofomata [2] can be classified into three relief units. These units are the plains and lowlands including all the river valleys, the cuesta landscapes and the highlands. It is observed that the uplands which are made up of highly friable sandstones yield easily to erosion and induce gullying even on slopes of about 5%. The cuestas and other highlands with somewhat stable lithology resist gullying but provide aggressive runoff which moves down to devastate the lowland areas especially at the toe slopes and river head-waters. The popular or infamous Agulu-Nanka gully erosion sites started from the head waters of streams and slopes of Awka-Orlu Upland region. The genesis and location of this particular gully site on the landscape is similar to numerous other gully sites in the region.

Figure 2. Typical gully site

4. Influence of climate

The rainfall of southern Nigeria generally is heavy and aggressive. Rainfall amount ranges from over 2500 mm in the southernmost region towards the Atlantic Ocean to about 1500 mm annually around River Benue in the northern borders. Rainfall intensities are high and often above 50 mm/h with short interval intensities in excess of 100 mm/h. Rainfall often come between the month of March and last till October. In some years the rainy period is unduly prolonged while in other years their onset may be delayed for more than 5 weeks. The present global climate change resulting from El-Niño and has not helped issues in this regard.

The nature of the rainfall regime contributes significantly to the erosivity of rainfall. Rainfall erosivity is the potential ability of rain to cause erosion. It is also a function of the physical characteristics of rainfall. Obi and Salako [11] reported that the raindrop sizes obtained generally in the Guinea savannah ecological zone of West Africa ranged from 0.6 to 3.4 mm. The mean drop sizes (D_{50}) of 28 rainfall events ranged from 1-1 to 2.9 mm. There are experimental evidence to suggest that intensity and energy are likely to be closely linked with erosivity. A number of statistical relations have been established in the past between the erosive power and amount of rainfall in other parts of the tropical region [12, 13, 10, 14]. The best estimator of soil loss was found to be a compound parameter, the product of the kinetic energy of the storm and intensity. In Nigeria, the total kinetic energy load of 1091 mm rainfall

http://dx.doi.org/10.5772/51020

at Samaru in Northern Nigeria was about 3600 Jm^{-2}. This was twice the amount recorded in southern Africa by Stocking [15]. However, the product of the kinetic energy of the storm and the maximum intensity of the rainfall during the first 30 mins of a storm (EI_{30}) was most significantly correlated with soil loss determined on standard field plots [16]. Erosivity values therefore have been used successfully to produce iso-erodent map of West Africa [14].

In southeastern Nigeria, Obi and Ngwu [17] characterised the rainfall regime and recommended Lal's index of Aim as having advantage over other indices of erosivity such as KE> 1 and EI_{30}. However, Salako et al. [18] compared all the available indices of erosivity adopted in southeastern Nigeria and came up with some modifications of existing ones. Two indices E_kI_{30} and E_kI_m were recommended

Where;

E_k is in $MJha^{-1}$ (kinetic energy)

I_{30} is 30 minutes rainfall intensity and

I_m being maximum intensity computed over a 6 minutes duration

Rainfall therefore plays very significant roles in the erosion hazard of southeastern Nigeria. The rainfall distribution, amount and intensity in combination of other environmental factors contribute in accelerating the rate of interrill rill and gully erosion in southeastern Nigeria. This is evidenced in the sense that as rainfall amount decrease northwards, the rate of all types of soil erosion by water decreases.

5. The influence of vegetation

The constant deforestation of the former rainforest due to population explosion and increased agricultural activities in the region expose the bare soils to the vagaries of weather thus escalating the soil erosion problems. The implication is that the soils are frequently subject to different degrees of erosion including accelerated erosion. Vegetation and land use are one of the most important factors in soil erosion process in southeastern Nigeria. Stocking [15] noted that vegetation acts in a variety of ways by intercepting raindrops through encouraging greater infiltration of water and through increasing surface soil organic matter and thereby reducing soil erodibility. According to Lal [19], choosing an appropriate land use can drastically curtail soil erosion.

In southeastern Nigeria soil erosion especially gullies are most intensive on soil on which the former growth has been disturbed, that is mostly on agricultural soils stripped of growth for reasons of infrastructural developments such as road and housing construction. Ofomata [3] showed that in the region soil erosion is connected mainly with agricultural activities and other related land use activities such as mining, road building, urbanization, industrialization and general infrastructural development. These land use activities deprive the soil surface of its vegetation and also contribute directly to sliding, slumping, interrill and rill erosion including gullying.

6. The influence of geology

The general influence of lithology on soil erosion processes is manifest directly by the resistance of the denuded bed rocks exposed to the flow of water and affected by the character of parent materials whose properties are given by the bed rock. The direct effect of bedrock is also manifest in the properties of the soil forming parent materials which conditions the principal properties. Some geological materials are vulnerable than others to aggressive energy of the rainfall and runoff. High erosion risks match with units of weak unconsolidated geological formations. This is more pronounced when such geological units coincide with medium to long and even very long slopes with marked gradients.

In Nigeria, Ofomata [1] classified the potential erosion susceptible areas based on underlying geology. He indicated that areas of high susceptibility correspond to geological regions of weak unconsolidated sandy formations while least susceptible areas are within the consolidated tertiary to recent sediments. Also in southeastern Nigeria, the classical gully sites are located in the False-bedded sandstone, Coastal Plain sands, Nanka Sands and the Bende-Ameki Formations. These are all sandy formations which have more gullies than their Shale formation counterparts. In these formations, there exist the sites of worst catastrophic soil erosion in the whole of sub-Saharan Africa. The geology therefore plays direct and indirect influence on the gully formation. The indirect effect is on the soil formation and the nature of soil which contribute significantly to erosion processes. The influence of soil process on soil erosion often referred to as erodibility is the subject of discussion in the next section.

7. The influence of soil factor (erodibility)

The erodibility of the soil is defined as the vulnerability or susceptibility of the soil to erosion. It is a measure of a soil's susceptibility to particle detachment and transport by agents of erosion. Igwe [20] remarked that a number of factors such as the physical and the chemical properties of the soil influence erodibility. In southeastern Nigeria, the nature and the long weathering history of the soils parent material evident in the dominance of the clay mineralogy by non-expanding minerals and low soil organic matter concentration due to high mineralization rates and excessive leaching of nutrients could be linked to the worsening situation. The highly weathered soils contain high concentrations of Fe and Al oxides. Inappropriate land use and soil management options are also a common feature of agriculture in the region. Anthropogenic factors often combine to weakened soils to produce severe gullies. The soils are hence loose and slumps under high intensive rainfall that renders them easily detachable. Some of the soils have the tendency to slake and form seals under such intense rainstorms thereby resulting in considerable runoff and soil erosion. The soil erodibility factor has since been recognized as a contributing factor to soil erosion hazard. The erodibility of the soils in terms of soil indices that predict or promote soil erosion will be elaborated on. The contributions of soil factors to soil erosion in Nigeria have variously been discussed [21, 20, 22]. Igwe et al. [21] found that the soil

http://dx.doi.org/10.5772/51020

clay content, level of soil organic matter (SOM) and sesquioxides such as Al and Fe oxides, clay dispersion ratio (CDR), mean-weight diameter (MWD) and geometric-mean weight diameter (GMD) of soil aggregates all influence soil erosion hazards in southeastern Nigeria. SOM, Al and Fe oxides control dispersion and flocculation of the soils. In the event of very aggressive rainfall, the soil inherent properties often combine with the physical forces of rainfall to produce soil erosion in the soils.

Figure 3. Gully cutting

Erodibility varies with soil texture, aggregate stability, SOM contents and hydraulic properties of the soil. Igwe [22] claimed that the soil dispersion ratio (DR) and the clay dispersion ratio were good indices of erodibility. The soils with high water-dispersible clay (WDC) in southeastern Nigeria often create problem in that in tilled land use, mud flow and soil loss from runoff cause major alteration in the stream flow within watersheds causing severe environmental challenges. Soil crusting, sealing resulting from aggregate breakdown are secondary problems arising from deposited sediments. The large particle sizes are resistant to transport because of the greater forces required to entrain these large particles while the fine particles are resistant to detachment because of their cohesiveness. Aggregate stability and associated indices have been shown to be most efficient soil properties that predict the extent of soil erosion.

In other parts of the world the use of aggregate stability indices in predicting soil erodibility have shown reliable information on the extent and degree of soil erosion [23, 24]. In Western Europe, Le Bissonnais [25] indicated that the mean-weight diameter (MWD) of soil aggregates was a very reliable soil property that could show the erosion potential of the soil in the sense that MWD predicts soil erodibility. Therefore aggregate stability and MWD are very reliable properties in explaining, quantifying or predicting soil erosion and other soil problems such as crusting and sealing.

Again other soil properties encourage structural failure, sliding and mass movement of soils. These soil factors are the mineralogy of the clay and even the soil chemical properties. The stability of the soil mass is therefore depended on the clay minerals present. Illite and smectite more readily form aggregates but the more open lattice structure of these minerals and the greater swelling and shrinkage which occur on wetting and drying render the aggregates less stable than those formed from kaolinite. Soils in which either kaolinite or illite clay predominates but contains small amounts of smectite are easily dispersive. Smectitic soils are more erodible than the soils that contain only small amount of smectite. Conversely, soils that do not contain smectite are more stable, less erodible and less susceptible to seal formation.

Figure 4. Gully site in association with interill and rill erosion

The sodium dithionite extractable Fe oxide is a soil chemical property which relates significantly with erodibility of the soil. This particular property affects the soil structure and the soil fabric, often being responsible for the formation of soil aggregates and cementation with other major soil components [26, 27]. The mechanism of aggregation of soils in southeastern Nigeria in the presence of Fe (Hydr) oxide has been demonstrated [8, 26, 27]. The presence of OH-Al polymers may lead to a reduction in the swelling and expansion of clay particles by bonding adjacent silica sheets together and by displacing interlayer cations of high hydration power and thus promoting aggregation. Well crystallized aluminium hydroxide may also be able to act as cementing agent in acid soils such as in southeastern Nigeria but its magnitude may be negligible as compared with non-crystalline materials. Iron oxides

http://dx.doi.org/10.5772/51020

therefore are more effective than aluminium hydroxide in cement effectiveness except for soils undergoing frequent oxidation-reduction processes.

8. Anthropogenic influence

An important factor which contributes significantly to soil erosion problem in southern Nigeria is anthopogenic influence arising from misuse of land. Poor farming systems have contributed to collapse of soil structure and thus encouraging accelerated runoff and soil loss due to erosion. In the event of uncontrollable grazing caused by the nomads has resulted in deforestation of the landscape while indiscriminate foot paths created on the landscape has helped the incipient channels on the landscape to form. These channels eventually metamorphose to gullies especially when they are not checked at inception. Road constructions including uncontrolled infrastructural developments have contributed significantly in gully developments. Some road networks under construction have been abandoned in the region due to gully formation.

Figure 5. Gully about cutting an asphalt surfaced road

9. Identification of gullies and erosion sites

Soil erosion sites in southeastern Nigeria have been identified through various methods. In the 1960s and 1970s gullies were enumerated through natural resource surveys but this method proved to be very cumbersome and often do not actually represent the actual situation on ground. This led to the use of aerial photo interpretation (API) in the generation of information for soil erosion studies. Niger Techno & Technital Spa [28] employed API in the documentation and publication of soil erosion problems in eastern Nigeria. The other methods in this category is remotely acquired data from satellites, radar imageries and from geographic information systems GIS. The advantages of API and other remotely acquired information is that the information they show are real and exact and sometimes in real time. However, acquiring information through this source is very expensive and most often unaffordable by some governments and establishments in Nigeria.

Of late modelling soil erosion hazard in southeastern Nigeria has been very useful not only for erosion hazard prediction but for conservation purposes. Ofomata [3] used multiple regression equation with the environmental factors of climate, vegetation, soil and anthropogenic factors being the variables to predict the soil erosion hazard in southern Nigeria. Again, Igwe et al. [21, 8] employed two different models in predicting soil erosion hazard in some parts of southeastern Nigeria. The predictive abilities of these models with some modifications were satisfactory and approximated data obtained from the field. Table 1 presents the predictive ability of some soil parameters in 25 selected soils in the region.

Soils	Soil Classification	DR	CDI	RUSLE K	CFI	MWD	GMD	Overall* Ranking
Oseakwa	Gleysols	5	10	24	10	10	13	11
Akili ozizor	Fluvisols	21	22	22	22	18	16	24
Osamala	Cambisols	17	13	22	13	1	4	10
Oroma etiti	Gleysols	15	25	21	25	2	5	20
Umuewelu	Cambisols	21	20	19	20	3	6	18
Ezillo	Acrisols	16	12	13	12	17	12	17
Abakaliki	Acrisols	10	16	13	16	7	2	8
Okija	Ferralsols	9	8	8	8	4	9	4
Ogurugu	Fluvisols	11	17	11	17	15	8	13
Nenwe	Cambisols	14	19	25	19	5	9	19
Ifite ogwari	Cambisols	18	18	16	18	21	25	21
Umueje	Cambisols	18	7	16	7	23	7	12
Umumbo	Cambisols	24	15	16	15	8	3	14
Omasi	Acrisols	25	21	8	21	22	19	22
Adani	Acrisols	23	23	15	23	24	23	25
Nsukka	Acrisols	12	8	12	8	20	21	14
Obollo afor	Acrisols	12	2	5	2	12	15	5
Nteje	Ferralsols	3	6	5	6	15	20	7

Soils	Soil Classification	DR	CDI	RUSLE K	CFI	MWD	GMD	Overall* Ranking
Awka	Acrisols	2	1	5	1	13	16	2
Idodo	Acrisols	20	24	20	24	14	14	22
Ukehe	Acrisols	7	4	8	4	19	22	8
Abor	Acrisols	4	4	2	4	9	18	3
Nachi	Acrisols	6	11	4	11	25	24	14
Nanka	Acrisols	7	14	2	14	11	1	6
Nawfija	Acrisols	1	3	1	3	5	11	1

*1- most erodible; 25- least erodible; DR- Dispersion ratio; CDI- clay dispersion index; RUSLE K- Wischmeier erodibility factor (K); CFI- clay flocculation index; MWD- mean-weight diameter of soil aggregates; GMD- geometric mean-weight diameter of soil aggregates (Source: Igwe et al [21]).

Table 1. Ranking of soils in order of significant erosion predictability

In other parts of the world the use of some soil parameters such as the water-dispersible clay (WDC) has been adopted as a major parameter in soil erosion models as in the Water Erosion Prediction Project (WEPP) [29]. This method has been widely used in the development of soil erosion models for some parts of eastern Nigeria [22]. Soils with high WDC have high soil erosion potential and therefore WDC constitutes a great problem to the soil and the entire environment. The negative influence of high clay dispersion on soil erosion results in detachment, transportation and deposition of sediments with essential plant nutrient elements down stream. This clay associated sediments constitute high environmental menace to man, livestock and agricultural fields. The streams and rivers are silted, while the aquatic life suffers serious problems due to high concentration of nitrates, organic matter and phosphorus in clay suspension down stream. These information have served as basic information for soil conservation processes

10. Control and Remediation

The state of soil erosion problem in southeastern Nigeria calls for a comprehensive soil conservation programme so as to check catastrophic erosion hazard. The soil conservation measures should be those farming system practises which ensure sustainable soil productivity while maintaining equilibrium between the ecosystem and regular anthropogenic influence. In the design of soil conservation strategies, the permissible soil loss tolerance so as to avoid catastrophe in the event of failures of such strategies. In the United States of America, the permissible soil erosion loss is between 2.5 and 12.5 t ha^{-1} y^{-1} [7], while in Czech Republic, Holy [30] noted that the permissible soil loss was between 1.0 to 16 t ha^{-1} y^{-1} in very deep soil of 120 cm thickness. Obi [31] observed that for a highly weathered, porous and deep ultisol in southeastern Nigeria, the tolerable soil loss was about 10 t ha^{-1} y^{-1} under maize production, with appreciable loss in the production capacity of the soils.

Therefore, the suggested soil conservation measures based on the agricultural land use is recommended for the entire agro-ecological system. The land use option suitable to the area should be that based on integrated watershed management with arable farming, agroforestry and intensive afforestation. These practises are considered cheap option which can be afforded by the rural poor farmer. The methods are also very sustainable and not destructive to the agricultural land. This is aimed at reducing the annual soil loss rate and prevents the development of fresh gullies in the area. Agricultural land use should be based on topographic variations, major soil distribution, soil potential erosion hazard, hydrology and other geomorphological variables. Igwe [4] recommended that the entire region should be partitioned into 4 broad sections based on their location on the landscape. The lowlands and valley floors which also contain sediments should be put to rainfed and irrigated farming of arable crops. The main soil conservation strategies should be those that improve plant nutrient availability, land levelling in case of irrigation and drainage. On the land areas that are on 5% slope and below, the regular recommended cultural practises of organic matter application to the soil is suggested while mulching, crop rotation and well managed agro-forestry are some of the ways of keeping the soil uneroded. Crop residues in association with tillage systems contribute immensely in the conservation of the soil or other wise. The other remaining 2 land units are those that vary between 5-30 % slope and mostly the sites of catastrophic gullies in the area. They should be permanently forested and may be used for wildlife conservation. The kind of forestation should be that which produces intimate multistoreyed association of woody species, grasses and creeping legumes. This will ensure steady cover for the bare soil and offer some kind of protection to the soil against the high intensive and aggressive rainfall. The major soil conservation strategies are broad-based terraces and cover cropping of bare soils. A more comprehensive soil conservation method will involve the application of certain hydrological or bioenvironmental processes so as to control the overland flow and excessive runoff.

11. Conclusion

Soil erosion in the form of gullies is very common in southeastern Nigeria. This review has shown the influence of geology, climate, geomorphology (slope), vegetation, man and soil itself on gully development and soil erosion in general. Typical empirical examples are cited from previous works from other researchers in other parts of the world and locally. Past works on estimation of potential soil erosion hazard in the region indicate that more than 1.6% of the entire land area has been devastated by gullies. The inherent characteristics of the local soils to a large extent promote the spread of soil erosion especially the gully type in the region. The roles of anthropogenic factors with regards to land use and its influence on the vegetation are considered. The serious deforestation of the vegetation and poor revegetation or afforestation programmes have all contributed to the catastrophic erosion hazards. General strategies for soil conservation with respect to soil erosion should include a more comprehensive soil conservation method which will involve the application of certain hydrological or bioenvironmental processes so as to control the overland flow and excessive runoff.

Author details

C.A. Igwe

Address all correspondence to: charigwe1@hotmail.com

Department of Soil Science, University of Nigeria, Nsukka, Nigeria

References

[1] Ofomata, G.E.K. (1981). Acrtual and Potential erosion in Nigeria and measures for control. *Soil Science Society of Nigeria Special Monograph 1*, 151-165.

[2] Ofomata, G. E. K. (1975). Soil erosion. *Nigeria in maps, Eastern States*, Ethiope Publishing House, Benin City Nigeria.

[3] Ofomata, G. E. K. Soil erosion. Southeastern Nigeria: the view of a geomorphologist, Inaugural lecture series University of Nigeria Nsukka (1985).

[4] Igwe, C. A. (1999). Land use and soil conservation strategies for potentially highly erodible soils of central-eastern Nigeria. *Land Degradation Development* [10], 425-434.

[5] Igwe, C.A. (1994). The applicability of SLEMSA and USLE erosion models on soils of southeastern Nigeria,. *PhD Thesis*, University of Nigeria, Nsukka.

[6] Giordano, A., Bonfils, P., & Briggs, D. J. (1991). Menezes de Sequeira E., Roquero D.L.C., Yassoglou A. the methodological approach to soil erosion and important land resources evaluation of the European community. *Soil Technology* [4], 65-77.

[7] Renard, K. G., Foster, G. R., Weesies, G. A., Mc Cool, D. K., & Yoder, D. C. (1997). Predicting Soil Erosion by Water: A Guide to Conservation Planning with the Revised Universal Soil Loss Equation. *U.S. Department of Agriculture, Agriculture Handbook*, 703, 384pp.

[8] Igwe, C. A., Akamigbo, F. O. R., & Mbagwu, J. S. C. (1999). Chemical and mineralogical properties of soils in southeastern Nigeria in relation to aggregate stability. *Geoderma* [92], 111-123.

[9] Hudson, N.W. (1981). *Soil conservation*, Cornell University Press, New York.

[10] Lal, R. (1976a). Soil erosion on alfisols in Western Nigeria. I. Effects of slope, crop rotation and residue management. *Geoderma*, 16, 363-373.

[11] Obi, M.E., & Salako, F.K. (1995). Rainfall parameters influencing erosivity in southeastern Nigeria. Catena. 24, 275-287.

[12] Morgan, R.P.C. (1974). Estimating regional variations in soil erosion hazard in Peninsular Malaysia. *Malaysian Nature Journal* [28], 94-106.

[13] Kowal, J. M., & Kossam, A. H. (1976). Energy load and instantaneous intensity of rainstorms at Samaru, Northern Nigeria. *Tropical Agriculture* [53], 185-198.

[14] Roose, E.J. (1977). Application of universal soil loss equation of Wischmeier & Smith in west Africa. *Greenland D.J. & Lal R (eds), Soil conservation and management in the humid tropics*, 177-188.

[15] Stocking, M.A.A. (1987). Methodology for erosion hazard mapping of the SADCC region. *Paper presented at the workshop on erosion hazard mapping, Lusaka, Zambia, April.*

[16] Lal, R. (1976b). Soil erosion on alfisols in Western Nigeria. I. Effects of rainfall characteristics. *Geoderma*, 16, 389-401.

[17] Obi, M. E., & Ngwu, O. E. (1988). Characterization of the rainfall regime for the prediction of surface runoff and soilloss in southeastern Nigeria. *Beitrage für Tropicalische Landwirtschaften und Vetrinarimedizin* [26], 39-46.

[18] Salako, F. K., Obi, M. E., & Lal, R. (1991). Comparative assessment of several rainfall erosivity indices in southern Nigeria. *Soil Technology* [4], 93-97.

[19] Lal, R. (1983). Soil erosion in the humid tropics with particular reference to agricultural land development and soil management. *Proceedings of the Hamburg Symposium. IAHS Publication 140, August*, 221-239.

[20] Igwe, C.A. (2003). Erodibility of soils of the upper rainforest zone, southeastern Nigeria. *Land Degradation & Development* [14], 323-334.

[21] Igwe, C. A., Akamigbo, F. O. R., & Mbagwu, J. S. C. (1995). The use of some soil aggregate indices to assess potential soil loss in soils of Southeastern Nigeria. *International Agrophysics* [9], 95-100.

[22] Igwe, C.A. Erodibility in relation to water-dispersible clay for some soils of eastern Nigeria. Land Degradation and Development (2005). , 2005(16), 87-96.

[23] Bryan, R.B. (1968). The development, use and efficiency of indices of soil erodibility. *Geoderma* [2], 5-26.

[24] Bajracharya, R. M., Elliot, W. J., & Lal, R. (1992). Interrill erodibility of some Ohio soils based on field rainfall simulation. *Soil Science Society of America Journal* [56], 267-272.

[25] Le Bissonnais, Y. Experimental study and modelling of soil surface crusting processes. *Catena Suppl.*, 1990(17), 13-28.

[26] Igwe, C. A., & Stahr, K. Water-stable aggregates of flooded inceptisols from south-eastern Nigeria in relation to mineralogy and chemical properties. Australian Journal of Soil Research (2004). (42), 171-179.

[27] Igwe, C. A., Zarei, M., & Stahr, K. (2009). Colloidal stability in some tropical soils of southeastern Nigeria as affected by iron and aluminium oxides. *Catena,* 2009(77), 232-237.

[28] Niger Techno & Technital Spa. Pre-feasibility study of soil erosion in East Central State of Nigeria (1979). pp., 1

[29] Lane, L.J., & Nearing, M.A. (1989). USDA-Water erosion Prediction project. *Hillslope profile version. NSERL Report 2 West Lafayette,* 259.

[30] Holy, M. (1980). *Erosion and environment- Environmental Sciences and applications,* 9, Pergamon Press Ltd, 225.

[31] Obi, M.E. (1982). Runoff and soil loss from an oxisol in southeastern Nigeria under various management practices. *Agricultural Water Management* [5], 193-203.

Chapter 9

Daily Flow Simulation Using Wetspa Model with Emphasize on Soil Erosion (Study Area: The Neka Catchment in Mazandaran Province, Northern Iran)

Ali Haghizadeh

Additional information is available at the end of the chapter

http://dx.doi.org/10.5772/51420

1. Introduction

Soil erosion by water is one of the most important land degradation processes in Mediterranean environments. This process is strongly linked to problems of flooding and channel management. The relationship between land use and erosion in mountainous forested watersheds has been known in a qualitative sense for some time. Vegetation management, forest road construction, and forest fires, impact basin sediment yield by increasing the amount of sediment available for transport and the amount of surface water available to transport it.For early flood warnings as well to get time for planning and operation of civil protection measures it has become very important that forecasts are made and simulation of floods is carried out. With the ever increasing demand for water resources, it has become very important that the natural processes of floods be predicted, so that current and future environmental issues can be addressed well in time. A simplified representation of the natural hydrological system is the hydrological model. In this model, different physical processes are represented at different time scales and at a wide range of times. This has basically been associated with a lack of appropriate observational data to constrain model states, increase in the number of model outputs [18] and lastly, the complexity of the model. Basically, the distributed hydrological models give us the opportunity to deal with forcing implement these models.The models will help provide the means by which important information regarding existing and future stream flow conditions, very important information regarding hydrological state variables and the state of knowledge on basins of interest can easily be captured for use. Every entity in Iran has suffered great losses due o the floods together with socio-economic development. Among all the different kinds of natural disasters, flood is ranked first in terms of frequency, affected area, losses caused and the severity. On August 10th, 2001, a big flood took place in Golestan and Gorgan River with return period of 200 years and caused a lot of damage. The damages caused by the flood include 15,000 hectares damages to agricultural lands, 10,000 people rendered homeless, 10,000 hectares of damage to forests, and the greatest loss was the human death toll. 247 human beings had been killed in the disaster. The total damage to the entire province was a staggering 491 billion Rials. Once again on July 29th, 1999 Neka city in Mazandaran province was hit by a flood similar to the one which had hit the Golestan and Gorgan Rivers. About a billion dollar worth of damage was caused, including more than 4000 shops and homes damaged to about 50% to 100 %, 400 km railway damaged. 33km of road and about 100 people injured. The Neka river basin is located in northern Iran. It is frequently affected by storms and heavy rain which causes inundation. Flood forecasting modelling is the most important component of the real-time flood forecasting system. This system can mitigate such natural disasters. Flood warning and forecasting systems mostly use hydrologic/hydraulic models. These models, when optimally validated and calibrated can be very effective in minimizing flood damage through non-structural means. In the early years, these flood models were very simple with sophistication in technology comes effectiveness. With advances in geographic information systems and remote sensing theses models have now become more effective. The advantage of these models is that spatially distributed basin characteristics on stream flow can be reflected by these models. There are various studies in which this particular model has been applied, including the Alzette river basin in Luxembourg [9], Barebeek catchment in Belgium [3], the Hornad watershed in Slovakia [2], the Suoimuoi catchment in northwest Vietnam [8], the Simiyu river (Lake Victoria) in Tanzania [16] and the Suriname river basin, in central Suriname [14].

2. Methodology

2.1. WETSPA model

The WETSPA model which was proposed by Wang et al, (1996) [19] and predicts regional or basin Water and Energy Transfer between Soil, Plants and Atmosphere. The unique thing about it is that it has a physical basis and is a distributed hydrological model, laying down the concept of the composition of a basin hydrological system from atmosphere, canopy, root zone, transmission zone, and saturation zone. In order to briefly describe the model, it is vital to mention that heterogeneity in the basin is dealt with by its division into various grid cells which have further divisions for maintaining water and energy balance into a bare soil and vegetated portions. Another feature is that a vertical flow in a single dimension simplifies the movement of water in the soil and is inclusive of surface infiltration percolation and increase in capillary in the area which not saturated and underground water's recharge. All this can be done by defining the proper and appropriate parameter classes of land use, topography and soil type, and also by using the available data base.

The model for root zone water balance for each grid cell is obtained by Inputs and outputs equations:

$$D\frac{d\theta}{dt} = P - I - S - E - R - R \quad (1)$$

where D (m) is root depth, θ (L^3L^{-3}] is soil moisture, t (m) is time, I (LT^{-1}) is initial abstraction including interception and depression losses, S (LT^{-1}) is surface runoff or rainfall excess, E (LT^{-1}) is evapotranspiration, R (LT^{-1}) is percolation out of the root zone, and F (ms^{-1}) is interflow. The assessment for excess rainfall is done by means of modification in a moisture-related rational process with a latent runoff coefficient with due consideration to factors such as land cover, slope, soil type, magnitude of rainfall, and pre soil moisture.

$$S = c\left(P - I\right)\left(\frac{\theta}{\theta_s}\right)^{\alpha} \quad (2)$$

where θ_s(L^3L^{-3}) is soil porosity or saturated water content, *c* (-) is potential runoff coefficient.A lookup table has been used for deriving the values of *C*, with associated values to slope, soil type and land use classes [10]. Literature was searched to obtain default values but they may be changed by the used if necessary as suitable according to a region's specific situation. Liu.et al (2005) [8] did this by creating a look up table for catchments, associating latent rainfall excess coefficient to various arrangements of slope, soil type and land use. This study utilizes these values. Rainfall excess coefficient, given by :

$$\alpha = \max\left[1, K_{run} + \left(\frac{1 - K_{run}}{P_{\max}}\right)\right] \quad (3)$$

Where K_{run} (-) is surface runoff exponent and P_{max} (LT^{-1}) is a rainfall intensity scaling factor. Is impacted by rainfall intensity and the effect is reflected by α (–). *A* is greater when the rainfall intensity is not high and as a result, surface runoff is lower, and the approach is shifted towards high rainfall intensity leading to runoff and soil moisture's linear association. When the surface runoff exponent is 1, it implies P_{max} parameter, which is threshold rainfall intensity, leading to linearity between actual runoff coefficient and comparative soil moisture content. In order to measure the scale for this parameter, we can compare observed and computed peak discharges when floods are high. The calculation of evapotranspiration from soil and vegetation is done on the basis of a relationship explored by Thornthwaite and Mather (1955) [17] as defined by probable growth level, evapotranspiration, vegetation type, and soil moisture content:

$$\begin{cases} E = 0 & for \ \theta \prec \theta_w \\ E = \left[c_v K_{ep} E_p - I \right] \left[\dfrac{\theta - \theta_w}{\theta_f - \theta_w} \right] & for \, \theta_w \le \theta \le \theta_f \\ E = c_v K_{ep} E_P - I & for \, \theta \ge \theta_f \end{cases} \tag{4}$$

Where c_v (-) is vegetation coefficient which varies throughout the year depending on growing stage and vegetation type, K_{ep} (-) is a correction factor for adjusting potential evaporation E_p (LT^{-1}),θ_w (L^3L^{-3}) is moisture content at permanent wilting point, and θ_f (L^3L^{-3})is moisture content at field capacity. Only increase in groundwater capillary can bring about evapotranspiration when wilting point ($\theta < \theta_w$) is higher than the water content, and groundwater storage *G*(m) and a scaling parameter *Go*(m) controls groundwater capillary rise:

$$E_g = (c\ K_{ep} E_p \text{ - } I\ )\ G / G_o \qquad for \ \ q < q_w \tag{5}$$

Where E_g (LT^{-1}) is the evaporation from groundwater, K_{ep} (-) is a correction factor for adjusting potential evaporation E_p (LT^{-1}). Pan measurement or Pemman-Monteith or other equations utilizing accessible weather data, referring to water surface or a grass cover in large fields provide the model's latent evaporation E_p but the actual latent evaporation may be defined by native aspects that these methods do not attend to. There is need of correction factor K_{ep} (-) for calculating these effects and on average, the value is somewhere around 1, and the model can measure this through a water balance simulation over the long term. Various flows are obtained using water balance equation. List includes merely interflow, percolation, groundwater flow and excess rainfall as these components were adding to stream flow as the aim was its simulation. The model evaluates overland flow and channel flow routes by a linear diffusive wave estimate of the St. Venant momentum equation where the equation models the cell's flow process as [12, 15]:

$$\frac{\partial Q}{\partial t} + c_i \frac{\partial Q}{\partial x} - d_i \frac{\partial^2 Q}{\partial x^2} = 0 \tag{6}$$

Where Q (m^3/s) is the flow discharge at time t (s) and location x (m), c_i is the kinematic wave celerity at cell i (m/s), d_i is the dispersion coefficient at cell i (m^2/s). Manning relation can estimate the parameters c_i and d_i, which are required for defining the function of cell response, as [5]:

$$c_i = \frac{5}{3} v_i \tag{7}$$

$$d_i = \frac{v_i R_i}{2S_i} \tag{8}$$

Where R_i is the average hydraulic radius or average flow depth of cell i (m), S_i is the cell slope (m/m), and v_i is the flow velocity of the cell i (m/s) calculated by the manning equation. De Smedt F. et al. (2000) [3] and Liu et al. (2002, 2003) proposed an amateur passage time distribution, an estimated numerical key to the equation of diffusive wave associating the discharge when flow path concludes to the initial accessible runoff.

$$U_i(t) = \frac{1}{\sqrt{2\pi\sigma_i^2 t^3 / t_i^3}} \exp\left[-\frac{(t-t_i)^2}{2\sigma_i^2 t / t_i} \right] \tag{9}$$

Where t_i is the mean flow time from the input cell to the flow path end (s), and $\sigma_i{}^2$ is the variation of the flow time (s^2).There is spatial distribution for t_i and $\sigma_i{}^2$, flow celerity and dispersion coefficient functions define it through convolution integral and flow paths given by topography:

$$t_i = \sum_{j=1}^{N} \left(\frac{1}{c_j} \right) l_j \tag{10}$$

$$\sigma_i^2 = \sum_{j=1}^{N} \left(\frac{2d_j}{c_j^3} \right) l_j \tag{11}$$

Surface water moves faster than groundwater and therefore the latter is made easy in the form of a lumped linear reservoir on sub-catchment scale, derived from GIS. Sub-catchment outlet's groundwater flow is connected to overland flow and interflow that is routed to the primary channel from every cell, keeping in view the impact of river damping on every part of the flow. This is followed by routing the total hydrograph to the basin outlet by Eq. 6 derived channel response functions. The sum of the discharge is attained by convoluting every cell's flow response. Simulation of every hydrological process within GIS is a benefit of this methodology.

Calibration process includes simulated discharge's comparison with respect to observed discharge after the model has been executed. Model calibration is done after adjustments of input parameters and evaluation of the output. The input parameters being used were: groundwater recession coefficient (Kg), scaling factor for interflow computation (Ki), temperature degree-day coefficient (K_snow), initial groundwater storage (G_0), rainfall degree-day coefficient (K_rain), rainfall intensity corresponding to a surface runoff exponent of 1 (P_max), surface runoff exponent for a near zero rainfall intensity (K_run), and correction factor for potential evapotransipiration (K_ep). For minimization of the differences between simulated and observed discharge, graphical technique was used. Certain parameters were highly influential as far as simulated flow is concerned. These included groundwater recession coefficient, scaling factor for interflow computation, maximum groundwater storage, and initial groundwater storage. When observed flow was in excess of the simulated flow, there was an increase in the initial groundwater storage. When simulated discharge was in excess of the observed discharge, there was a reduction in the interflow-scaling factor. Manual adjustments in the parameters of the model were done to achieve a good match between outlet's observed and simulated flow.

A statistics series is used for evaluation of observed hydrograph reproduced by WetSpa. Various factors are taken into account, including model confidence, evaluation on the basis of visual comparison, model efficiency, the bias, time to the peak, and evaluation of peak flow rate. Statistical data provides quantitative estimations of goodness of fit between predicted and observed values. These also indicate the extent to which observations reflect predictions. Assessment of the model's predictive capabilities is done on the basis of test's outcomes. Evaluation of the goodness of fit in the time to peak or the peak discharge may be done with the help of its absolute or relative errors, and the other criteria for evaluation are explained below:

1) Model Bias

Model bias is defined to be the relative mean difference between observed stream flows and predicted stream flows for a simulation sample that is sufficiently large. It should also be reflective of the ability to reproduce water balance.

$$MB = \frac{\sum_{i=1}^{N}(Qs_i - Qo_i)}{\sum_{i=1}^{N} Qo_i} \tag{12}$$

where CR1 is the model bias, Q_{si} and Q_{oi} are the simulated and observed stream flows at time step i (m^3/s), and N is the number of time steps over the simulation period.The equation 2.12 shows the criterion. It basically gives systematic over-prediction or under-prediction for prediction sets. The fit is considered better when the MB value is low. Observed flow volume's perfect simulation is represented with the value 0.0.

2) Modified Version of Nash-Sutcliffe Efficiency for High Flow Assessment

Equation 13 describes a slightly different version of the Nash-Sutcliffe criterion. Actually, it is a fusion of Hoffmann, El Idrissi et al et al calibration condition (2004) [7] :

$$NSH = 1 - \frac{\sum_{i=1}^{N}\left(Qo_i + \overline{Qo}\right)\left(Qs_i - Qo_i\right)^2}{\sum_{i=1}^{N}\left(Qo_i + \overline{Qo}\right)\left(Qo_i - \overline{Qo}\right)^2} \tag{13}$$

Where NSH is a modified version of the Nash-Sutcliffe criterion for gauging the aptitude with which the time evolution of high flows has been simulated. The formula shows how high discharges bear greater load as compared to low discharges. The value of 1 is the optimal value in NSH.

3) The Modified Correlation Coefficient (r_{mod})

$$r_{\mathrm{mod}} = \left[\frac{\min\{\sigma_o, \sigma_s\}}{\max\{\sigma_o, \sigma_s\}} * r\right] \tag{14}$$

Where$\sigma_0$and$\sigma_s$are the standard deviations of observed and simulated discharges respectively, r is the correlation coefficient between observed and simulated hydrographs.The Modified Correlation Coefficient (r_{mod}) is for the purpose of judging the accuracy of the replication of time progression of high flows. 1 is the perfect value for r. Model calibration allows one to discover the most suitable values for global modern parameters. The numerical information incorporated in the study is derived from Model Bias (MB), indicating the accuracy of water balance simulation, the Modified Correlation Coefficient (r_{mod}), which shows divergences in hydrograph shape and size [11], and the modified Nash-Sutcliffe efficiency for high flow (NSH), which weighs up the simulation of the stream flow hydrograph (13) provided by the functions below. The Aggregated Measure (AM) below is brought in to assess the efficacy of model workings during the calibration and confirmation phases (Eq 15). It will compute various features of the reproduced hydrograph such as shape, size and volume:

$$AM = \frac{r_{\mathrm{mod}} + Ns + (1 - MB)}{3} \tag{15}$$

AM requires a value of 1 in order to produce complete correlation. The time periods listed in Table 2 have been used in order to better sort out the proficiency of the model's workings. (1; 6).

This includes the trial-and-error approach, the computerized method of numerical parameter optimization, or a mixture of the two. The PEST program was implemented for the process of design auto-calibration (4). There is a possibility of happening upon a local optimum instead of its global counterpart because the optimization subroutine is a local search technique. Clearly, in order to get suitable initial parameter values, a primary manual calibration is needed. The PEST program is used to regulate the WetSpa model with the obtained initial parameters values. One of the primary functions of the program is the forecasting of floods at the basin opening, and so we assume that the high flow reports are more significant than the low flow numbers in the model calibration routine.

Category	Aggregated measure(AM)
Excellent	>0.85
very good	0.70-0.85
good	0.55--0.70
poor	0.4-0.55
very poor	<0.40

Table 1. Model performance categories to indicate the goodness fit.

2.2. Study area

The Neka basin is situated in Northern Iran and river is the central branch of Neka River. The Ablu station is situated next to the convergence of the Neka River and watershed covers the area of 1864.74km^2 up to the Ablu station. The Neka Basin is a large catchment with heights ranging from 36m to 3814m. Mean elevation is 1531m, mean slope is 24.81%. The Max flow length is 163km and stream order outlet is 133. Other characteristics are given in Table 1 and Figure 1.

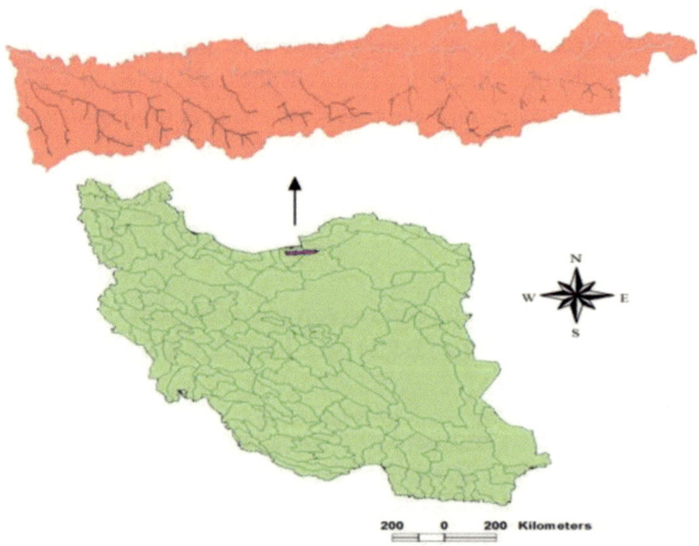

Figure 1. Location of the neka watershed in Iran.

A digital elevation model (DEM) was extracted from the topography map supplied by Iranian National Geographical Organization (IRNGO). The map was transformed into a 50m grid DEM.

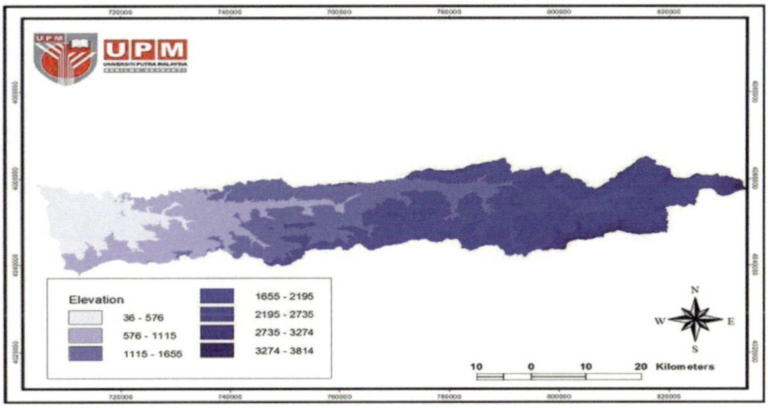

Figure 2. Elevation map of Neka basin.

Iranian Soil Conservation and Watershed Management Research Institute of Iran (SCWMRI) provided the data for ground cover (Figure 3) depicts the land use map for the study, presenting land cover for year 2000. 5 types of land cover can be observed, 45% of the basin is covered by forest land, 40% by rangeland, agriculture and villages account for 3% and the last 12% by short grass (figure3).The soil data was not available for the study, thus, soil map was obtained using a resource evaluation and land capability map (1973) from the Iranian Agriculture and Natural Resources Ministry, Soil sciences and Fertility Institute. 3 different types of soil are present in the catchment. Clay loam accounts for 89.4% of the basin, 5.6% is silt loam and the remaining 4.8% is sandy loam (Figure 3).

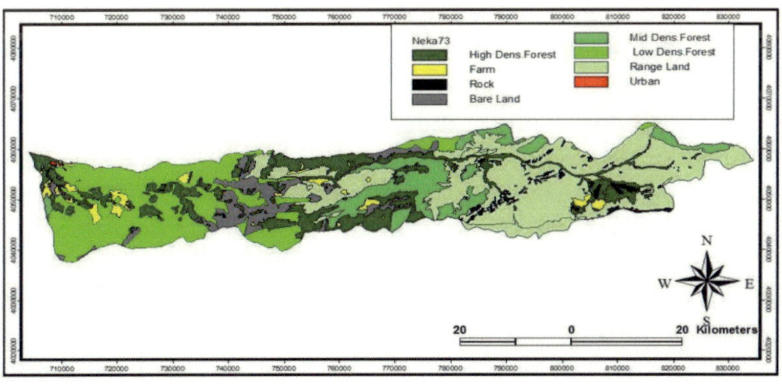

Figure 3. Land use types in Neka basin.

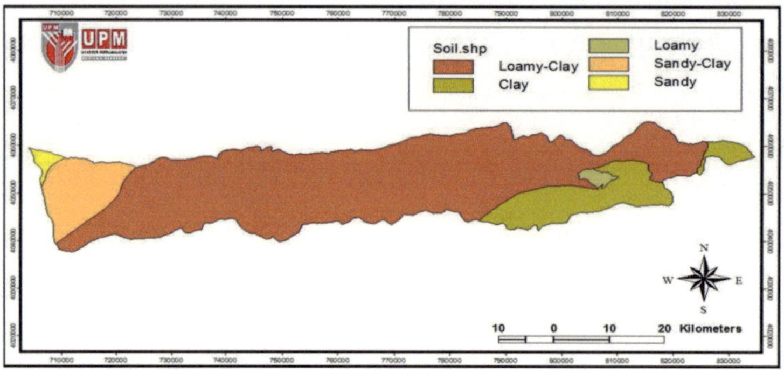

Figure 4. Soil types in Neka basin.

2.3. Inputs

For this study, precipitation, the potential evapotranspiration (PET) and discharge data were obtained from Water resources management of Iran(WRM.IR). The data included daily discharge data at 1 gauging station, precipitation for 7 stations and PET for 4 stations. All the data sets covered a period of 14 years i.e.1986-1990 and 2002-2004 Though daily discharge data is available for some locations in the catchment, only the Ablu station and the Neka station are being used for the study and model calibration.

2.4. Stream flow simulation by WetSpa model in Neka basin

Identification of spatial model parameters starts once the data is collected and processed for the use in the WetSpa model. Territorial features are taken out from the DEM including flow direction and accumulation, stream network, link and order, slope, elevation and hydraulic radius. To define the stream network the threshold is set to 10 i.e. when the upstream drained area exceeds 0.1km^2 then the cell is considered drained. To establish sub-catchments the threshold value is set at 1000, through which 265 sub-catchments were found with an average area of 6.971km^2.DEM extracted the slope map. The two are comparable with level slope (0.005 to 17.603%) in the middle area where as steeper slopes (17.603 to 158.389 %) were observed along the borders. The small area along the border has a very high gradient showing a ridge. Flow within the catchment is defined by the slope. In the above case the flow will be from the borders to the middle area of the catchment. The threshold value of 0.005% is considered for minimum slope while creating the grid of the surface slope. Should the calculated value be lower than the threshold value then to evade the inert water and low speeds, the value is taken as 0.005 %.If the two similar streams order come together then the stream order rises by one. It keeps on increasing as the river runs its course since more and more streams join it. Stream order ranges from 1 to 133 in this catchment.100 cells were chosen to classify first order stream. Meaning the first order stream comes into existence if the runoff from 100 cells provide to one cell. If one first order stream meets another one then they form a second stream order.Hydraulic radius was extracted from DEM (flow accumulation). Power law relationship with greater probability was employed which explains and uses the relationship between controlling area and hydraulic radius. It ranges between 0.009m for upland to the outlet of main channel at 6.098m in this catchment. With the progressing of river at the downstream the hydraulic radius is escalating, which is very obvious. Hydraulic radius is minute as the flow of water in the upstream area of the catchment is not fast. Beyond the rate of 0.5 (2-year return period). The hydraulic radius framework is produced. Subsequently, porosity, residual moisture; plant wilting point, field capacity, the grids of soil hydraulic conductivity and pore size distribution index are categorized on the basis of the soil texture grid with the help of attribute lookup Table 2.4.The first moisture map was developed with the help of soil map. Largely, in the region of catchments the moisture map differs as of 0.1-0.6 but, it usually has the high value of about 1 in the flow of the river. To measure the overspill from catchments the moisture map is used. The areas where the humidity or moisture is more, the overspill water is high as well. With the help of soil map the soil hydraulic conductivity was developed. In the flow of river, the difference of hydraulic conductivity is 0.6-1.51, in the majority regions of catchments. The map is utilized to find out the overspill from catchments.Another map developed with the help of soil map is the soil porosity. The difference of hydraulic porosity is 0.43-4.75 in the course of stream.Other things developed or derived from soil map include pore size distribution index, field capacity, Neka catchment, residual moisture and plant wilting point. The runoff coefficient is developed from the slope whereas land and soil type utilize maps. The runoff coefficient varies from 0.071 to 1 in the above catchment; conversely the runoff coefficient in the majority regions of the catchment is from 0.071 and 0.587. In the middle region of the catchment, the runoff coefficient is optimal because the land is used for cultivation and agriculture and due to the presence of clay soil. Because of the occurrence of steepness in the lofty regions, the overspill/runoff coefficient is highest. In the same way Manning's roughness n coefficient, the grids of root depth and interception storage capacity are categorized again from land, making use of the grid. Manning's n for channels is interposed on the basis of stream order grid/framework in which 0.050 $m^{-1/3}s$ is set for lowest order, whereas 0.030 $m^{-1/3}s$ is set for highest order (Figure 5).

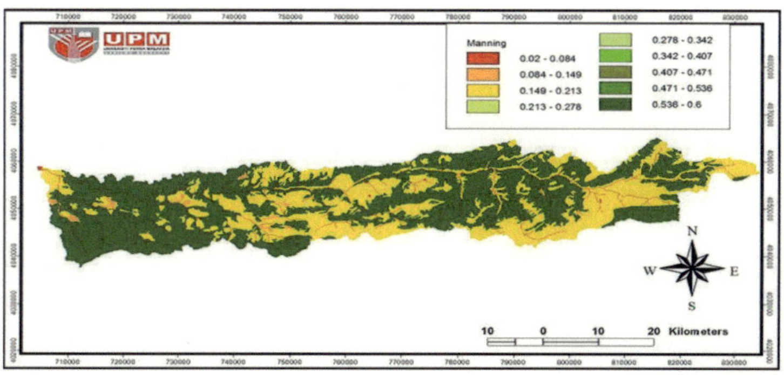

Figure 5. Manning coefficients map of Neka basin.

On the basis of hydraulic radius and Manning's coefficient, the velocity map was formed. The Manning's coefficient was acquired with the help of land use while hydraulic radius and the slope were developed from DEM. Manning's equation helps in calculating velocity (Equation 16).

$$v_i = \frac{1}{n_i} R_i^{2/3} S_i^{1/2} \tag{16}$$

Where, n_i is the Manning's roughness coefficient (m^3/s), R_i is the average hydraulic radius of cell i (m), S_i is the cell slope (m/m), and v_i is the flow velocity of the cell i (m/s).Velocity map shows the high velocity/speed regions (around m^3/s) from the river/stream course. Besides the vicinity of high stream order, the low speed/velocity is found in catchments.The travel time map notifies the flow time of water in the catchments that differs from 0-54.91 hour. In the catchment inlet the travel time is optimal while in catchment outlet its zero. With the help of time deviation factor, the velocity of water is measured. The standard deviations are produced as a result of the movement of the flow velocity from time to a basin exit/outlet. It allows computing the IUH from every grid cell near the basin outlet. The projected standard flow time is shown in the Figure 14 between the grid cells and the basin outlet.Grids of depression storage capacity and potential runoff coefficient are attained with the help of attribute tables, which are formed by joining the grids of soil, elevation and the land use. 3 per cent is set as a proportion of impermeable region in villages. For the whole catchment the computed standard potential runoff coefficient is of 0.85. The gridirons for temperature, PET and precipitation are formed on the basis of environmental or geographical coordinates of every measuring location besides this on the catchment frontier utilizing the Thiessen polygon extension of the ArcView Spatial Analyst (Figures 6 and 7).

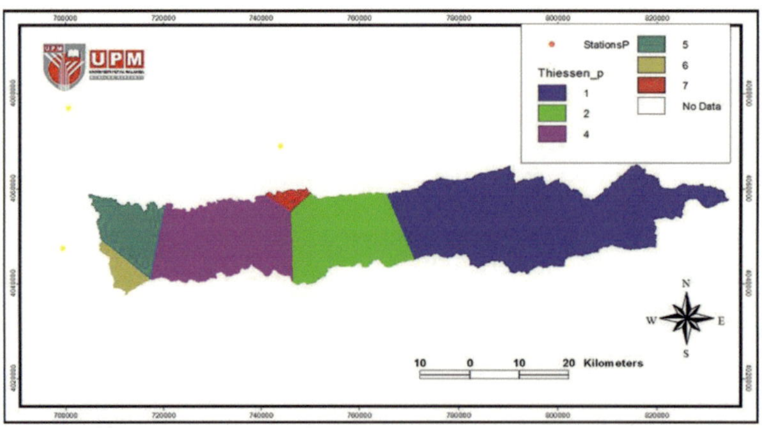

Figure 6. Thiessen polygons for the precipitation stations.

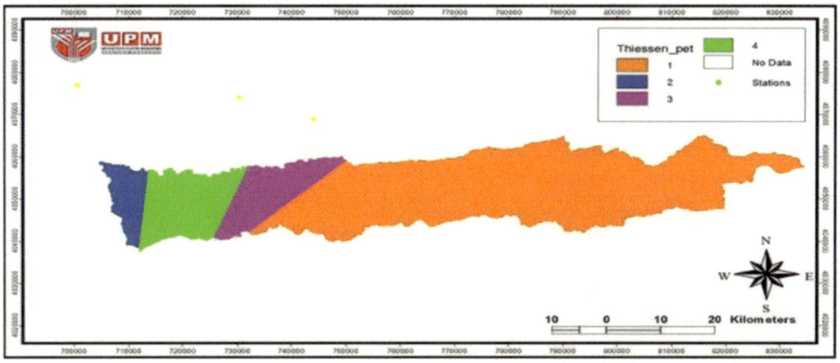

Figure 7. Thiessen polygons for the evaporation stations.

3. Results and evaluation

Forming validation and calibration, the 14 years (1986-1999) and (2002-2004) calculated PET, discharge data and the daily precipitation. From the 14 years time, the initial 12 years are selected for model calibration whereas the last two years are selected for model validation. Two models are used; the first one is the global model that is for the calibration processes while the second model being the spatial model whose factors remain unchanged. Global model factors are exclusively selected on the basis of basin traits, as talked about in the user manual of the model and the concerned documents. The imitation outcomes are evaluated to the experimental hydrograph on the Ablu station in the catchment basin statistically as well as graphically. By examining the base flow that is estranged from observed hydrograph, the first groundwater flow recession coefficient is anticipated. Amendments are mandatory of the parameters with respect to the total flow capacity and the fitting of base flow. The interflow scaling aspect is attuned for the recession and peak area of the flood hydrograph, which is receptive on behalf of high and low flows. Additionally, there are two parameters that are managing the quantity of surface runoff, which include the surface runoff exponent of about zero rainfall power and the rainfall intensity equivalent to the surface runoff exponent of about 1. These are attuned mostly for little storms as the real runoff coefficient is less because of the low rainfall intensity. With the evaluation of the water balance and hydrographs for the initial period, the active groundwater storage and soil water are attuned. The optimal active groundwater storage manages the quantity of vapour emerged from the groundwater and as a result it can be attuned by evaluating the flow amount in dry stages. As of the last two years of 14 year period the calibrated global parameters are utilized to reproduce the daily stream flow (Table 2).

No	Symbol	Parameter	Feasible range	Estimate	PEST (95%confidence interval)
1	Ki	Interflow scaling factor	0-10	1.263	1.2-1.3
2	kg	Groundwater recession coefficient (d-1)	0-0.05	0.0001	0.0079-0.0001
3	K_ss	Initial soil moisture (mm)	0-2	0.8424	0.98-1
4	K_ep	Correction factor for PET (-)	0-2	0.5373	0.53-0.54
5	G0	Initial active groundwater storage(mm)	0-500	1	1-2
6	G_max	Maximum active groundwater storage(mm)	0-2000	50	42.5-59.4
7	K_run	Moisture or surface runoff exponent (-)	0-8	7.8573	7.5-8
8	P_max	Maximum rainfall intensity (m)	0-500	133.9	130-250
9	T0	Threshold melt temperature (°C)	-1-1	0	0-0.53
10	K_snow	Melt-rate factor (mm $°C^{-1}d^{-1}$)	-2 - 2	-2	-2-0.5
11	K_rain	Rainfall melt-rate factor ($°C^{-1}d^{-1}$)	0-0.05	0	0-0.0211

Table 2. Calibrated model global parameters.

The model is calibrated using daily stream flows observations for Neka basin. Calibration is first done manually (trial-Error method) and then automatically using PEST to minimize the sum of square differences between observed and predicted stream flow. The validation as well as calibration periods found the performance of the model satisfactory. Tables 2 and 3 give the criteria for evaluation for the periods of calibration and validation. Hoffman et al. (2004)(7) gave four evaluation criteria, which are used here. The observations of high flow values are going to be more important in the model calibration than the low flow values because prediction of floods at the basin outlet is one of the primary purposes of the model. The evaluation in the process is carried out according to the criteria of Nash-Sutcliffe efficiency for high flow evaluation and the modified Correlation Coefficient r(mod) and Aggregated Measure(AM).The respective values were found for the validation period: the Nash-Sutcliffe efficiency was equal to 0.87, the flow volume was found to be 1.3% underestimated, whereas the modified Correlation Coefficient was 0.747 and Aggregated Measure was found to be 0.85 that performance is excellent for validation phase,Also The respective values were found for the calibration period: the Nash-Sutcliffe efficiency was equal to 0.78, the flow volume was found to be 2.2% underestimated, whereas the modified Correlation Coefficient was 0.734 and Aggregated Measure was found to be 0.83 that performance is very good for calibration phase. These results show that a lot of factors including the precipitation, antecedent moisture, and runoff generating processes were considered by the model in a manner that was spatially realistic and the basis for it was topography, land use and soil type. These factors resulted in producing highly accurate rates for the capture of both high flows as well as the general hydrological trends (Table 3).

http://dx.doi.org/10.5772/51420

Period	MB(CR1)	NSH(CR5)	r(mod)	AM	Performance
23/9/1986-31/12/1998	0.094	0.76	0.733	0.82	very good
(calibration period)	0.041	0.74	0.729	0.81	very good
	0.118	0.75	0.725	0.78	very good
	0.018	0.78	0.734	0.83	very good
22/9/2002-21/9/2004 (validation period)	0.012	0.87	0.746	0.85	Excellent

Table 3. Model performance of WetSpa model for calibration and validation period.

Simulated flows for the whole amount of time (1986-1999) and (2002-2004). Figures 8 shows a Graphical comparison between observed and simulated river discharge in Neka basin at 2002-2004. In general, one can notice a reasonable agreement between model results and observations. Peaks in the hydrograph are rather well predicted. As well for size as for time of occurrence.

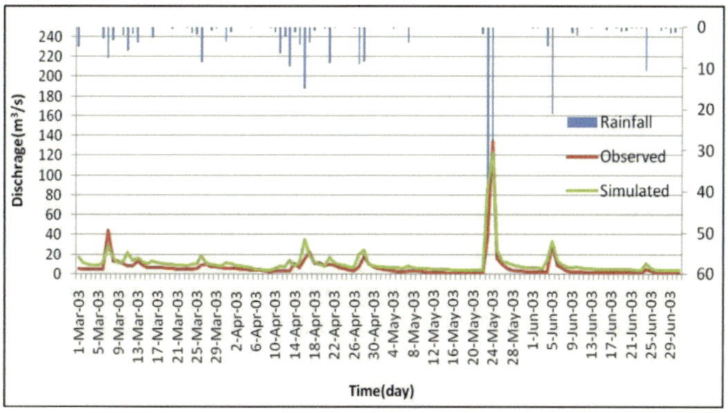

Figure 8. Observed and simulated daily flow at Neka basin for validation phase.

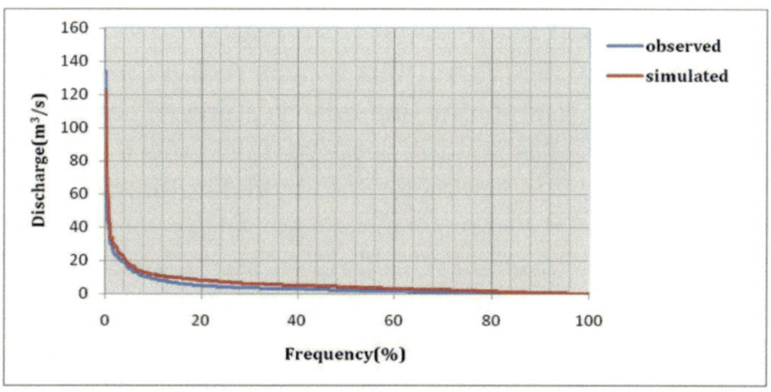

Figure 9. observed and simulated flow versu ferqunecy.

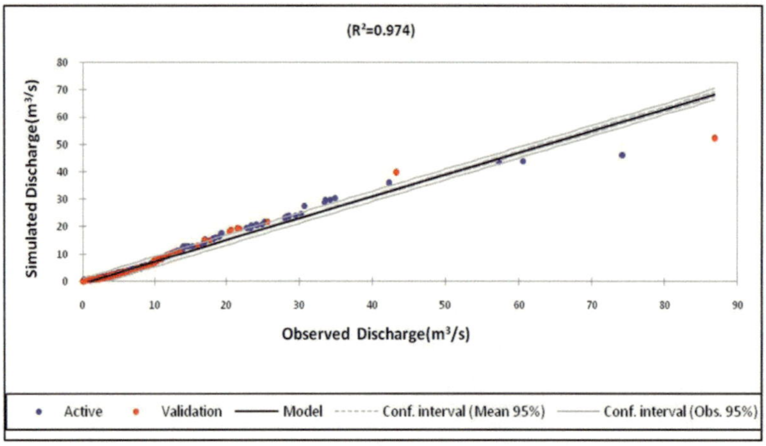

Figure 10. scatter plot of observed versus simulated flow for validation phase.

4. Conclusion and recommendation

This research allowed for the model to be examined for the Neka catchment in Iran for a period of 14 years. The daily rainfall and evaporation input data was tested. A fine comparison with the calculated hydrograph was obtained. The precision of this reproduction is around 0.87, 0.974 and 0.747 according to the Nash-Sutcliffe criteria, square correlations, and modified correlation coefficient respectively. The aggregated measure (AM) was found to be 0.85 that the model performance is excellent for the validation phase. This shows that the model is suitable for using rainfall data, antecedent moisture and evapotranspiration etc and generating runoff in a spatially rational based on the topography, land use and soil type and creating a nominally higher precision simulation catchment for the high flows. The graphical comparison amongst observed and estimated daily flows for the 14 years of model simulation proves that the season flood hydrographs have been properly reproduced by the distributed hydrological WETSPA model and this is achieved via reaching conclusions from using parameters based on the topography and other features of basin such as soil and land use. The consistency of the model estimation depends on how well the model"s construction is created along with how well it is parameterized. Model calibration is also essential to fix the model"s workings (8). Manual and mechanical calibrations are the two kinds of parameter estimations which are used. Mechanical calibration uses a search algorithm to check the perfect parameters and allows for numerous benefits using a physical approach. WETSPA model parameterization and the spatial format of the model parameters are estimated by employing the accessible field information to define the most essential differences. This approach ensures that the model uses that data that has been represented in the catchment. In this research, the WETSPA model was first manually calibrated through a trial and error process more using more than 500 tests for finding best parameters. This approach also allows for the use of automatic calibration methods to enhance the working of the model. The research recommendations are:

- An improved investigation can also presented the impact of coordinated alterations to the weather and land use on the hydrological procedures within the region, such as impact on flooding and soil moisture distribution.
- Using practical procedures to allow for the growth which accounts for a joint effect of elevation, slope, general movement of the atmosphere etc on the spatial division of rainfall, temperature and PET. This can also lead to an increased consistency of the model inputs and lower the indecision of the model outputs. This is important when designing for a large mountainous catchment. The radar data can also be added to the WetSpa model to simulate the spatial division of rainfall at each step at a time.
- Some of the model errors are caused due to lack of accurate and efficient input data such as rainfall, temperature and evapotranspiration. Hence, for increasing the efficiency of the model, increased number of rainfall, temperature and evapotranspiration stations, as well suitable distribution of the measuring stations over the watershed is required.

Author details

Ali Haghizadeh*

Address all correspondence to: haghizadeh.a@lu.ac.ir

Department of Range & Watershed Management Engineering, Faculty of Agriculture Engineering, Lorestan University,khorramabad, Iran

References

[1] Andersen, J., Refsgaard, J. C., & Jensen, K. H. (2001). Distributed hydrological modelling of the Senegal River Basin--model construction and validation. *Journal of Hydrology*, 247(3-4), 200-214.

[2] Bahremand, A., De Smedt, F., Corluy, J., Liu, Y. B., Poorova, J., Velcicka, L., & Kunikova, E. (2007). WetSpa Model Application for Assessing Reforestation Impacts on Floods in Margecany-Hornad Watershed, Slovakia. *Water Resources Management*, 21(8), 1373 -1391.

[3] De Smedt, F., Liu, Y. B., & Gebremeskel, S. (2000). Hydrologic modeling on a catchment scale using GIS and remote sensed land use information. *Risk Analysis II*, 295-304.

[4] Doherty, J., & Johnston, J. M. (2003). Methodologies for calibration and predictive analysis of a watershed model 1. Jawra. *Journal of the American Water Resources Association*, 39(2), 251-265.

[5] Henderson, F. M. (1966). Open Channel Flow. 522, Macmillan, New York.

[6] Henriksen, H. J., Troldborg, L., Nyegaard, P., Sonnenborg, T. O., Refsgaard, J. C., & Madsen, B. (2003). Methodology for construction, calibration and validation of a national hydrological model for Denmark. *Journal of Hydrology*, 280(1-4), 52-71.

[7] Hoffmann, L., El Idrissi, A., Pfister, L., Hingray, B., Guex, F., Musy, A., Humbert, J., Drogue, G., & Leviandier, T. (2004). Development of regionalized hydrological models in an area with short hydrological observation series. *River Research & Applications*, 20(3), 243-254.

[8] Liu, Y. B., Batelaan, O., Smedt, F. D., Huong, N. T., & Tam, V. T. (2005). Test of a distributed modelling approach to predict flood flows in the karst Suoimuoi catchment in Vietnam. *Environmental geology*, 48(7), 931 -940.

[9] Liu, Y. B., Gebremeskel, S., De Smedt, F., Hoffmann, L., & Pfister, L. (2006). Predicting storm runoff from different land-use classes using a geographical information system-based distributed model. *Hydrological processes*, 20(3), 533-548.

[10] Liu, Y. B., & Smedt, F. (2005). Flood modeling for complex terrain using GIS and remote sensed information. *Water resources management*, 19(5), 605-624.

[11] Mc Cuen, R. H., & Snyder, W. M. (1975). A proposed index for comparing hydrographs. *Water Resources Research*, 11(6), 1021-1024.

[12] Miller, W. A., & Cunge, J. A. (1975). Simplified equations of unsteady flow. Unsteady flow in open channels , 1, 183-257.

[13] Nash, J. E., & Sutcliffe, J. V. (1970). River flow forecasting through conceptual models part I--A discussion of principles. *Journal of hydrology,*, 10(3), 282-290.

[14] Nurmohamed, R., Naipal, S., & Becker, C. (2008). Changes and variation in the discharge regime of the Upper Suriname River Basin and its relationship with the tropical Pacific and Atlantic SST anomalies. *Hydrological Processes*, 22(11), 1650-1659.

[15] Olivera, F., & Maidment, D. (1999). Geographic information systems(GIS)-based spatially distributed model for runoff routing. *Water Resources Research*, 35(4), 1155-1164.

[16] Rwetabula, J., De Smedt, F., & Rebhun, M. (2007). Prediction of runoff and discharge in the Simiyu River (tributary of Lake Victoria, Tanzania) using the WetSpa model. *Hydrology and Earth System Sciences Discussions*, 4(2), 881-908.

[17] Thornthwaite, C. W., & Mather, J. R. (1955). The water balance. *New Jersey*.

[18] Wagener, T., & Gupta, H. V. (2005). Model identification for hydrological forecasting under uncertainty. *Stochastic Environmental Research and Risk Assessment*, 19(6), 378-387.

[19] Wang, Z. M., Batelaan, O., & De Smedt, F. (1996). A distributed model for water and energy transfer between soil, plants and atmosphere (WetSpa). *Physics and Chemistry of the Earth*, 21(3), 189-193.